Eddie Bauer

THE *Eddie Bauer* GUIDE TO FAMILY CAMPING

THE *Eddie Bauer* GUIDE TO FAMILY CAMPING

ARCHIE SATTERFIELD and EDDIE BAUER

ILLUSTRATIONS BY TED RAND

ADDISON-WESLEY PUBLISHING COMPANY
READING, MASSACHUSETTS • MENLO PARK, CALIFORNIA
LONDON • AMSTERDAM • DON MILLS, ONTARIO • SYDNEY

Copyright © 1982 by The Eddie Bauer
Company, Inc.

All rights reserved. No part of this publication
may be reproduced, stored in a retrieval system,
or transmitted, in any form or by any means, electronic, mechanical, photocopying, recording, or
otherwise, without the prior written permission of
the publisher. Printed in the United States of
America. Published simultaneously in Canada.

Library of Congress Cataloging in Publication Data

Satterfield, Archie.
The Eddie Bauer guide to family camping.

Includes index.
1. Camping—Equipment and supplies.
2. Camping.
3. Family recreation.
I. Title.
GV191.76.S27 1982   688.7'654   82-13923
ISBN 0-201-07776-0
ISBN 0-201-07777-9 (pbk.)

ABCDEFGHIJ-DO-85432

Design by Val Paul Taylor

INTRODUCING THE
Eddie Bauer
OUTDOOR LIBRARY

Eddie Bauer has been serving the needs of outdoor enthusiasts for three generations. Since 1920, we have been dedicated to developing, testing, and manufacturing the finest in apparel and gear for outdoor adventures. Our aim has been to make outdoor adventures more enjoyable.

Now, sixty-two years after we began, we are answering the needs of another generation of outdoor activists. *Eddie Bauer* Outdoor Guides like this one on Family Camping will help newcomers get started. Others will contain up-to-date, tested information to make your outdoor excursions safe, warm, dry, and comfortable.

CONTENTS

	PAGE
Acknowledgments	ix
Foreword by Eddie Bauer	xii

PART I — INTO THE GREAT OUTDOORS

Introduction		1
CHAPTER 1	CAMPING IN AMERICAN HISTORY	5
CHAPTER 2	ESPECIALLY FOR FIRST-TIMERS	15

PART II — GETTING READY

CHAPTER 3	CAMPING EQUIPMENT— THE ESSENTIALS	39
CHAPTER 4	TENTS AND SHELTERS	57
CHAPTER 5	BEDDING	71
CHAPTER 6	THE CAMPING WARDROBE	83

PART III — AT THE CAMPSITE

CHAPTER 7	SETTING UP CAMP	97
CHAPTER 8	THE OUTDOOR KITCHEN	115
CHAPTER 9	CAMP ACTIVITIES	143
CHAPTER 10	FIRST AID AND SAFETY	167
CHAPTER 11	STRETCHING THE SEASON	183
CHAPTER 12	WHERE TO NEXT?	191

PART IV — RESOURCES

CHAPTER 13	DESTINATIONS	203
CHAPTER 14	FURTHER READING	215
INDEX		219

ACKNOWLEDGMENTS

Books of this nature are always a team effort, and many individuals and companies have been very helpful in the research, reading drafts of the manuscript and bringing additional material to my attention.

Several members of the Eddie Bauer, Inc., staff have devoted many hours to this book, and it all began with a series of conversations with Jack Quinlan, assistant to the president. David V. Rudd, vice-president of marketing, has been the overall editor for the project, and much of the structure came from his suggestions. He in turn brought in other men and women from the company with their own areas of expertise. These include Jim Wheat, Ken Wherry, Bob Murphy, Marilyn Siehl, and Cort Green, recently retired from the staff.

Much of the material resulted from my own experiences camping with my family from the time the children were infants to the present, on trips all over the American West and in Alaska and the Yukon. Some of those trips were made through the cooperation of Kampgrounds of America (KOA) and Winnebago Industries, Inc.

For information on new developments in camping equipment, I have relied on the literature from Eddie Bauer, Inc., suppliers, and a selection of books, pamphlets, and product-information releases from them. Other material came from the following books:

Roughing It Easy (I and II) by Dian Thomas. New York: Warner Books, 1974, 1976.
Packrat Papers. Lynnwood, Washington. Signpost Publications, 1973, 1977.
Old Fashioned Dutch Oven Cookbook by Don Holm. Caldwell, Idaho: Caxton, 1969.
Medicine for Mountaineering by James A. Wilkerson, M.D. Seattle: The Mountaineers, 1967.
Map and Compass by Bjorn Kjellstrom. New York: Charles Scribner's Sons, 1976.
The Camping Trailer Handbook. Wichita, Kansas: Coleman Co., 1974.

One Pot Meals by Margaret Gin. San Francisco: 101 Productions, 1976.
The Picnic Gourmet by Joan Hemingway and Connie Maticich. New York: Random House, 1975, 1977.
Elegant Meals with Inexpensive Meats. San Francisco: Ortho Books, 1978.
Trailside Cookery by Russ Mohney. Harrisburg, Pennsylvania: Stackpole, 1976.
Campground and Trailer Park Guide. Chicago: Rand McNally and Co. Revised annually.
National Park Guide by Michael Frome. Chicago: Rand McNally and Co. Revised annually.
National Forest Guide by Len Hilts. Chicago: Rand McNally and Co., 1976.

And finally, four people were largely responsible for getting me involved in this projected series of the *Eddie Bauer* Outdoor Library: Richard V. Sawyer, who heard about it through the writers' grapevine and told me; Eddie Bauer, who was instrumental in my being selected as writer; Doe Coover, the Addison-Wesley senior editor, whose frequent trips to the Northwest have kept the book progressing professionally and whose format for the series has the simplicity of genius; and Dominick Abel, our literary agent, who fitted all the pieces together.

In spite of all this generous assistance from experts, some errors may find their way into the book like rain seeping through a poncho seam. If so, the fault is mine and not that of those who helped.

Archie Satterfield

FOREWORD

I am firmly convinced that one of the greatest benefits of wilderness camping as a form of recreation is that it promotes longevity. A dozen or so companions of my youth, both boys and girls, who joined me on wilderness treks have continued their keen interest in nature, have lived clean lives, and are now octogenarians like me and enjoy the same good health my wife, Christine, and I do.

During my fifty-five years of outdoor outfitting, I have always had a keen interest in those who traded with me, and I've been impressed with certain vital differences in their behavior.

Those who took advantage of every opportunity to get away from city life to enjoy outdoor treks—fishing, hunting, hiking, photography, or simply nature and clean outdoor living—have by and large enjoyed their retirement in good health.

In sharp contrast are those who told me, as they acquired costly outdoor equipment, that they were dedicated to first gaining financial security without taking time off for outdoor trips and waiting for their retirement years to enjoy outdoor recreation. Too many of these people soon died after retirement or suffered ill health, only to have their costly equipment brought back to me by their heirs for estate appraisal and usually disposal.

This is an example of what not to do. These people had denied themselves the enjoyment of the healthful outdoor life they had longed for, which would have prepared them for their retirement years.

I have been very fortunate in this respect. My business has also been my hobby, and when I opened my own shop it was like starting a lifelong vacation. I field-tested every piece of clothing and equipment that I sold so that people knew if it was in Eddie Bauer's store, he had personally put it to a rugged test. If I didn't trust equipment, it wasn't stocked. If I needed equipment that wasn't available elsewhere, I developed it myself.

This need for better equipment led to my invention of quilted goose-down clothing, followed by quilted goose-down sleeping bags. These inventions led to saving hundreds if not thousands of lives of armed forces personnel, including pilots and crewmen during World War II.

Throughout all the years I operated my own business, my employees and I were always out on camping trips, combining our recreation with field-testing equipment. If we said a good down coat would keep you warm in a storm on a glacier, we knew from personal experience that it would.

Family camping has been so rewarding to me and my many friends and customers that I'm confident this book will encourage other parents to take up family camping and introduce their children to this clean, healthful form of outdoor recreation.

Eddie Bauer

PART I
INTO THE GREAT OUTDOORS

INTRODUCTION

THE JOYS OF FAMILY CAMPING

If you have ever camped in the woods near a lake or stream, for the rest of your life you will remember the sound of the wind in the trees, the feel of the sun warm on your back, and the smell of coffee and bacon mingling with the woodsmoke. You will remember those evenings when the silence and stillness are punctuated only by the slap of a beaver's tail on the calm water and the fireflies trying to light up the night.

If you are lucky, you will wake before first light and lie in your cozy sleeping bag and watch the black sky slowly turn pale as the stars disappear, one by one, through false dawn into sunrise.

Although most campers speak of the adventures on camping trips, it is often the quiet little events you will remember for years to come. You may never discuss them because you consider these memories personal, but you will remember the smells of fresh leaves mingled with decaying wood, the patterns of sunlight filtering through the forest, the sounds of running water and the wind, birds calling and fish rising in the morning and evening.

The physical and mental benefits of camping are many, and few other forms of recreation offer more to strengthen the family bond. The distractions of young love, neighborhood bullies, stereophonic music, television, refrigerator raids, telephones ringing at dinnertime—all these recede once the family car is parked and you are setting up camp.

Camping offers peace and quiet far from the normal routine of our lives. It gives us an opportunity to get exercise in natural rather than urban, downtown-club settings. It puts us in a position of learning about nature firsthand rather than from books or television specials.

Children take to camping as naturally as they do to building sand castles on the beach. Household chores, a dreary form of undeserved punishment at home, are seen in a new light at the campground. Putting up tents and learning how to build fires and operate campstoves is great fun. Going to fetch a bucket of water isn't a chore as much as it is a small adventure.

It is a rare child who can't find something to do on a camping trip and to keep happily occupied almost every minute of the day. Infants will go anywhere their parents take them and will be happy if they're kept dry, warm, and well fed. Older children are so inventive and curious about new surroundings that they often resemble perpetual motion machines. A stick may be just a stick to an adult, but to a child it is a spear, a magic wand, a territorial boundary, a pencil for drawing pictures in the dirt, a cooking implement for hot dogs and marshmallows, a bow, or an arrow.

Perhaps more than any other family activity, camping adds pleasant experiences to a child's warehouse of memories, and happy adults are usually those who have pleasant memories from their childhood. These camping memories become an important part of an entire family's heritage, as much as the family scrapbooks children and grandchildren love to look through all their lives.

Just as world travelers bring back artwork, furnishings, and decorations from foreign countries, campers often make it a habit to bring home something to remind them of each trip. Some collect items for Christmas use, such as fir or pine cones to fashion into wreaths or to hang as ornaments. A handful of smooth river rocks that change colors when wet can be placed in a terrarium. Pieces of worm-carved wood and shells or sand dollars are prizes to be found on many beaches. Wild flowers can be pressed in a journal or diary with notes of where and when they were picked.

All these things, plus the simple pleasures of being outdoors in healthful surroundings, add to the family memories we cherish throughout our lives.

During recent years we have heard and read a great deal about the importance of taking up lifetime sports as opposed to the more strenuous team sports reserved for the young. Camping and its subsidiary activities are an ideal lifetime sport because they can be enjoyed throughout one's entire life, from infancy to old age. Camping can be as strenuous or as sedentary as you wish, from wilderness expeditions and mountain climbing to an overnight trip at a nearby campground.

This book, then, is a frank attempt to encourage more families to go camping. Although outdoor recreation has become one of the major leisure activities since the late 1940s, there still are millions of people who for various reasons have never spent a night in a tent. Except for strolls in city parks or along shores or nature trails in national parks, their experiences with nature have been limited to seeing it through their car windows.

Although being outdoors is a universal urge, many noncampers avoid it because they have had a bad experience on an earlier outing, or they are afraid they will have a bad experience. Another reason is the confusion over what kind of equipment is needed and the fear that acquiring this basic equipment requires an enormous investment.

Camping is a form of recreation that hardly knows social or racial barriers. Although you will frequently find the inevitable equipment snobs in some campgrounds, still it is a great social leveler, and you will find business tycoons and salaried wage earners shoulder to shoulder around the campfire.

Yet anyone can camp—young or experienced or neophyte. The out of doors isn't an alien environment to be feared. It is our natural habitat. And with today's equipment and camping conveniences, camping can be a pleasurable and beneficial experience for all. While it is true that campers run the risk of getting wet, chilled or otherwise uncomfortable, the proper clothing and other equipment will keep such problems to a minimum. Those who have gone on a few family camping trips learn to take such slight discomforts in stride and think of them as a reasonable trade-off for the pleasure of being outdoors and the closeness such trips almost invariably bring to families.

Because it is so popular, camping offers a wide variety of degrees of roughing it. You can go deep into the wilderness with everything you need in a backpack, or you can travel from campground to campground in a luxurious motorhome with all the amenities of home aboard. Between these extremes is a wide choice of how you enjoy the outdoors.

It is beyond doubt the most popular form of outdoor recreation in North America, and many people who enjoy outdoor trips don't really think of themselves as campers. They may be dedicated bird watchers or hunters or fishermen, and when they say they're going out on a trip, they seldom mention that camping is part of the program.

So this book has been written with the novice in mind, those who enjoy being outdoors but cannot find a book basic enough to provide answers to their questions. Since camping has become so enormously popular in recent years, the beginner has been ignored in favor of the more experienced people who know the basics. It is hoped this book will fill that gap.

CHAPTER 1

CAMPING IN AMERICAN HISTORY

The history of North America is rife with great adventures we would now call camping trips, wilderness treks or expeditions: Daniel Boone's steady westward push through Cumberland Gap to his final stop in Missouri; the great Canadian voyageurs who gradually pushed modern civilization across Canada to the Pacific and north to the islands of the Arctic Ocean; the farmers who floated their crops down the Mississippi to New Orleans and walked part of the way home on the Natchez Trace; the meat hunters who went out for days at a time to gather food for their families; the trappers and traders along the Ohio, Mississippi, and Missouri rivers.

The most dramatic examples of camping involved the great expeditions mounted during the nation's infancy when America was attempting to stretch its boundaries from the Atlantic to the Pacific. The best known of them all is the Lewis and Clark Expedition of 1804-6. No other undertaking was better planned or better carried out; they were gone two years, traveling in new and treacherous country with all the dangers of weather, terrain, disease, and unknown Indian tribes using equipment that today would make a dedicated camper shudder to contemplate. Yet when they returned to St. Louis from the mouth of the Columbia River, they had lost only one man, and he died of a ruptured appendix which would have killed him had he been in a Boston hospital at the time; the first successful surgery for appendicitis was still several years away.

The basic camping equipment

changed little from that period until the middle of this century. Firearms improved, boats and canoes had motors available, trains and automobiles enabled people to get into the wilderness quicker, but camping equipment was accepted as heavy and bulky.

But Americans have always been a restless people, and the European peasant who had his definite place in society back home came to America and found that he could let his imagination and ingenuity thrive. This freedom, the constant boom atmosphere in America, plus the most beautiful geography in the world kept Americans constantly interested in camping and made them seek ways to make it simpler and easier to enjoy. The outdoor life has always been one of the bonuses of living in America.

In many respects, camping in those early years was simpler because the population was so thin that camps could be set up almost anywhere. Those going some distance off the roads and railroads could use their team of horses and farm wagon or strings of pack animals to carry their equipment.

Some of the most famous campers during the late nineteenth century were the wealthy and members of European royalty who spent fortunes coming to America to visit the great wilderness west of the Mississippi. They went on safaris in Africa and expeditions in America, carrying every possible convenience with them, including gourmet cooks, servants, musical instruments, desks, bedroom sets, and vast collections of firearms.

For the rest, camping gradually became a form of recreation with short trips for hunting and fishing. Families would load up some kitchen equipment, a few blankets, some canvas, and put it all in the family wagon and go down to the nearby creek for a day or two of fishing and visiting with other friends on the same outing.

Sometimes itinerant preachers would come to a settlement and hold a revival meeting beside a stream where the converts could be baptized in a large group. Farm families often came to the area with their crude tents and fed the horses out of sacks of feed stored in the wagon.

Perhaps the greatest impact on camping as a form of recreation came from the establishment of the Boy Scouts in 1908 in Great Britain and two years later in the United States. The principles of the Boy Scouts that emphasize good physical and mental health, and lots of outdoor activities, were quickly adopted by Americans because the United States and Canada had more forms of outdoor recreation available than almost any other part of the world.

A major breakthrough in clothing and sleeping equipment came when Eddie Bauer invented the quilted goose down design for a coat. Not only is it pounds lighter than woolen and other insulating materials, the quilted-down coat he invented could be compressed into a tiny bundle for storage or carrying. When the coat was taken out and "spanked," the goose down sprang back to its original shape to give its original insulation qualities.

Shortly after this came an event that had an enormous impact on not only outdoor equipment, medicine, and plastics, but also the entire shape of the world. This event was World War II. Wars are filled with contradictions, of course, and one of them is that they kill and maim the young and innocent. They also speed up research and development of new medical and other equipment that improves the lives of those who survive the wars.

For campers, World War II meant that the use of down as insulation was seriously considered for clothing and sleeping bags. Since the silk supply was cut off by the war, industry developed synthetic replacements: nylon, rayon, and similar inexpensive yet lightweight, strong materials. The techniques and processes learned in the manufacture of parachutes and other equipment quickly found peacetime applications.

Some of those are the tents that will sleep six, yet can be carried by a child; parkas that are extremely durable, yet cost and weigh little; pack materials that also are lightweight and durable.

New processes for manufacturing aluminum were developed for airplanes and many other military uses. Those were soon translated into tent poles, pack frames, tent pegs, stove parts, and a multitude of other uses.

Thus, shortly after the war ended in 1945, America was ready for a revolution in outdoor recreation, and industry was frantically searching for peacetime products to manufacture. Before the end of the 1940s, America was camping in a big way. Those who could not find the proper equipment in sporting goods stores had the option of shopping at the thousands of military surplus stores that sprang up all over the country. Anyone who was alive during that period remembers with great fondness going to the surplus stores with their curious blend of odors of leather, mothballs, oils, preservatives, and new cloth. Those stores offered you parachutes still packed and ready for use, flight suits with down insulation, gas masks, belts, yards of webbing, command tents that would hold a dozen or more campers, miscellaneous spare parts for tanks, barrels of nuts and bolts, special tools that were good for absolutely nothing, steel helmets, crates of C- and K-rations, tail wheels from trainer planes, torpedo-shaped wing fuel tanks, and so forth.

Ingenious outdoorsmen could find uses for many of these products, and the garages and backyards of America were littered with surplus products that the families were going to get around to building one day.

When the war ended, another revolution of sorts began. Most Americans were better off financially than ever before, and Detroit went back into full production of cars and trucks. The great American urge to travel resumed, and many families whose first experiences with camping came during the Great Depression of the 1930s while they traveled America looking for work now began remembering those hard times with nostalgia. They remembered sleeping out beneath the stars in hayfields or along streams. Wouldn't it be fun, they reasoned, to do the same thing again, only this time by choice.

This urge for traveling, and America's great selection of places to see and things to do, almost naturally led to the next step in outdoor recreation—the recreational vehicle. This began as a trailer pulled behind the family automobile, not so much for recreation as for having a temporary home wherever the jobs were available. Americans had always slept in their vehicles, from the covered wagons that went across the plains and mountains to the Pacific Coast to the first trucks and pickups.

But now these vehicles became a luxury item. First the beds of pickups were enclosed with a simple box. Then people added ice chests, portable stoves, and gasoline lanterns and built racks to double as beds and storage areas. Then came the customized campers with natural-gas stoves and heaters, electric lights, holding tanks so indoor plumbing could be installed.

By the early 1950s, recreational

vehicle (RV) was a way of life in America. By the 1960s, franchised campgrounds for RVs were scattered all over America. The public agencies—National Park, National Forests, Bureau of Land Management, state and city agencies—all had to go into crash campground construction programs. America was going on a camping spree.

being used on some RVs to help them improve gasoline mileage. Aerodynamics helped modify RV designs so that they created less drag while being driven. New methods of waste disposal, such as using the heat from the exhaust system, have been more environmentally acceptable. New plastics have made all equipment stronger and lighter.

It has been ever since. Remote areas that had been visited by no more than 100 persons in all history suddenly were being visited by thousands of hikers and backpackers. Old trails that had not been used since the automobile arrived now were becoming gullies from all the footprints. Land that is of no use to farmers has been turned into commercial campgrounds. The banks of artificial lakes all across the continent are now lined with RVs and tents.

Camping equipment manufacturers are always on the lookout for ways to use new technology. As the space age required new materials, some of those found their way into camping circles, such as "space blankets," sturdy but extremely lightweight plastics that are

Obviously, camping is not a passing fad of a nation famous for its fads. It is a firmly entrenched way of life for many Americans, and no other form of family recreation comes close to its popularity. And it is still growing.

THE LITTLE CAMPERS

Adults often forget that children approach camping—the whole business of life—in a slightly different way from adults. Adults tend to think of camping as a serious, hard-nosed, man vs. nature activity. They also forget that children not only don't care about the Ten Essentials and calories and carbohydrates, but that most children don't even vaguely want to know.

Lecture them on these things all you want and the best you'll get from them is a polite silence while their agile minds flick from subject to subject. Look closely and you'll see a far-off look in their eyes.

This does not mean that children aren't eager to learn all the things adults know about camping; they would prefer on-the-job training rather than classroom settings.

So traveling with children in the outdoors means you have to meet them more or less on their own terms and not be disappointed when the trip ends without their having memorized the windchill chart or the capacity of a gas stove.

As mentioned elsewhere, infants and toddlers will go anywhere their parents take them and are usually manageable as long as they are dry, warm, and well fed. Of course all parents are subjected to the "terrible twos" when the ex-toddlers are amazingly contrary. But since their attention spans are still brief at the best, it is easy to bribe them into your good graces with a new game, a bit of tickling, a promise of a peanut-butter-and-jelly sandwich, or a "Look! There's a pileated woodpecker."

One of the most remarkable characteristics about children is their propensity for becoming totally exhausted almost as quickly as a bolt of lightning. One moment they are running pell-mell through the campground, and the next moment they are sound asleep on the cargo bag in which food is stored. A few minutes later they are awake again and going full speed as though nothing happened. This can be disquieting.

But that is the nature of children. They burn out fast, but their recuperative powers are staggering.

The instructional portion of camping should not be overemphasized. Children love to learn by doing, and learning to do the things adults do is part of the initiation into adulthood. There was a time not long ago (it still exists in some families) when labors were divided strictly according to gender. The menfolk did one kind of work; the womenfolk another. This isn't much of a problem in camping. So many men go off on hunting and fishing trips with their pals that the boundaries are blurred if not erased. Happily it is also relatively easy to get children to do at least a modest amount of work around the camp. Here, much more than at home, it is easier to sell the idea that those who work, eat; those who do not, go without.

Some camping cynics insist that camping and all the equipment it can entail is the ultimate in toys. Since a lot of camping gear isn't used very frequently at home, children tend to think of the stoves and tents and lanterns and sleeping bags as toys. Let them. They'll enjoy them much more if they are in a different category from light switches, electric ranges, refrigerators, and automatic washers.

Before the trip, go over the entire plan with the family. If the first day's drive is a long one, tell them so. But try to plan at least one stop every 100 or 150 miles so they can romp a bit. If you still have two hours' driving before you reach the goal and the youngsters ask if you're almost there, don't lie and say "almost," because they'll get even with you by grouching, whining, or poking each other.

One father likes to tell his children that they still have two hours to drive when they are actually less than two minutes from the campground. Early delivery on a promise is always pleasant.

One nice way to pass the time on a long car ride to camp is to require everyone to help tell a story. One of the parents will start an adventure story, build up to an exciting scene, then turn it over to someone else in the car. Require each member of the group to tell at least two minutes' worth at a time, or longer depending on the ages and inclinations of the group. This seldom works on the way home. Everyone seems to want to sleep all the way home.

When you go on day hikes or longer ones, prepare for all sorts of imagined cases of exhaustion. Children can always manage to have some kind of disaster that requires them to be carried part of the way, usually the last steps back to camp. It is a continual source of amazement to parents how calm children become when carried. After they have won this little victory, it is often easy for the bearer to pretend fatigue, too, and get the little con artist back on the trail.

By the time children are in the post-toddler period, around kindergarten or first grade, they want even more to be a part of the adult proceedings. This is a good time to get them a small pack. It doesn't have to be fancy or large enough for an expedition, but it should be large enough for them to carry their lunch or a sweater or something useful. Let them have a choice in colors or shape.

As soon as they are old enough to be trusted in a store without a leash or being carried, take them along on shopping trips for the camping trip. This not only is instructive; it also makes them feel a part of the whole expedition.

All the playground psychology you've learned at home comes into play when you arrive in a campground with other children. Pecking orders are quickly established among these total strangers. Like strange dogs getting together, children circle each other briefly and seem to know instinctively who is going to be the leader and who is going to be the follower.

One area where a very firm hand is needed is in the matter of campground good manners. Once people—all ages of them it seems—leave their hometowns, they seem to think they should be free to do as they please. Most adults manage to restrain themselves in campgrounds, but many children will take this opportunity to make more noise, break more glassware belonging to other campers, and tease more dogs than they have done in the past year at home. Unless they are exceptional or watched closely, all concepts of territorial boundaries are abandoned and the children will go tramping through other campers' territories. Since no fences exist and many campsites look alike, it is easy for them to confuse a campground with a city park without realizing that adults build invisible privacy areas around themselves wherever they go. This "personal space" is essential for everyone, especially adults, and children have to be taught its existence.

As children grow older and discover such amenities as hours-long telephone conversations and best friends, one way to enhance a camping trip is to permit each child to bring his or her best friend along. By this stage in their development children think parents are pretty boring people and siblings are wretched specimens of humanity. So the alternative to back-seat psychological warfare ("She's looking at me!") and campground brawls ("He put a leaf in my oatmeal!") is the presence of a best friend to give them something to do other than torment each other.

When you do take along a best friend, be certain to discuss your plans, menus, and equipment with the parents. Find out if the friend will eat anything other than graham crackers and pop.

CAMPING IN AMERICAN HISTORY 11

Children need to be part of the whole camping experience, including food preparation and cleanup.

Elsewhere in this book you will find the recommendation that each child be given a blank book, not a diary, to be used as a journal of camping trips. Parents can add to this storehouse of memories by taking photographs of the trips. Children love to look at family albums and will continue looking at them after they are adults. Few things bring more hysterical laughter among children than photographs of themselves when they were just kids. (They start saying that when they're around ten or twelve.)

Some families become addicted to movie film, which is certainly fine, but you can't decide on a moment's notice to go look at the camping photographs. A

With the continual improvement in amateur photographic equipment, almost everyone can take at least adequate photographs, and the camping camera does not have to be a sensitive, expensive camera. On the other hand, some campers take along the sturdy underwater cameras that are built for rugged use and can be used no matter the weather conditions.

Like adults, children have selective memories, and their lasting impression of a camping trip might be some embarrassing incident; a horrid little boy who dangled worms in people's faces, or how crabby Daddy was the last morning. But with a good selection of photographs to

good argument can be made for taking 35 millimeter slides and showing them on a projector after returning. The best or most memorable photos can be made into color prints for the family album.

jog memories, the unhappy moments soon blend into the whole scene.

Children are basically conservatives, if not reactionaries, at heart. If they had a great time at a campgrund last summer, they not only won't mind returning to it, they're likely to insist that you do. No matter how rosy the picture you paint for them of a new camping area, they approach it with the same apprehension they do moving to a new neighborhood. Parents simply have to live with this conservatism and continually introduce them to new adventures, new geography, and different types of camping trips.

The irony of this conservatism is that while they are afraid of a new piece of geography, they will immediately scamper up a leaning dead tree with total abandon. They help keep camping a lively form of recreation.

Much has been said and written about our obligations to the wilderness on behalf of our children and grandchildren, especially relating to wilderness preservation and expansion of lands set aside for the enjoyment of the people here now and those to come. You can tell children things like this, but since their idea of the future usually doesn't extend beyond tonight's marshmallow roast or tomorrow's sugar-coated cereal with powdered milk, don't expect them to be awed by the responsibility we feel for the outdoors. But they'll understand at about the age they decide parents are decent folk, flaws and all. And they'll appreciate your efforts.

CHAPTER 2

ESPECIALLY FOR FIRST-TIMERS

It is essential that a first camping outing for both children and adults be as pleasant as possible. If you plan an introductory outing and the weekend weather report indicates stormy or cold weather, it is best to postpone the trip. Later, as everyone becomes more experienced, camping in inclement weather often becomes part of the excitement of camping; it gives one a sense of accomplishment, and if you are properly outfitted, the discomforts are few.

Many campers, and books on the subject, make this relatively simple form of recreation seem, to the novice, only less complicated than launching a space shuttle. This is due in part to the wide variety of outdoor products on the market, and the many choices we have of places to go camping.

As with all forms of recreation, you can invest enormous sums of money in camping equipment, or you can shop carefully and have the best equipment at a more modest investment. Outdoor recreation has become very specialized, and it is possible, but not necessary, to feel compelled to buy three or four kinds of stoves, tents, and sleeping bags. With only a few exceptions, such as camping at extreme temperatures, you can buy a stove, tent, or sleeping bag that will be sufficient for all your camping needs for years and years.

Consequently, the emphasis in basic equipment is on quality so that most of your investments will be one-time expenditures. Under normal conditions, your camping equipment should last a lifetime with only occasional repairs. It isn't unusual for this basic equipment to last not only for years, but for generations so that it is handed down from parents to children like family heirlooms.

Few purchases of equipment are as versatile as those intended primarily for camping. Stoves and lanterns are a blessing in the case of power failures, as well as backyard cookouts, automobile breakdowns, and the like. Tents and sleeping bags are useful in slumber parties, and many families use the sleeping bags as comforters to keep winter heating bills down. The uses of outdoor equipment are limited only by the owner's imagination.

TYPES OF CAMPING

Camping comes in all manner of sizes and shapes, from overnights in your own backyard to several weeks in the wilderness with your home, kitchen, and bedroom on your back. If you are a first-timer who knows no more about camping than most of us do about the dead civilizations of Sumatra, here is a general breakdown of the forms from which to choose.

Backpacking: As stated earlier, this is the most rigorous form of camping because you carry your tent, food, cooking equipment, clothing, and various tools and medical supplies on your back. It has many advantages, primary among them the ability to go far from the crowds into wilderness areas. In addition to the equipment described throughout this book, you will need to add good boots and a pack to your equipment list.

Bicycle camping has become more and more popular during the past few years as people become more aware of the energy problem and equally interested in physical fitness. For this kind of traveling and camping, lightweight and compact equipment is as essential as it is for backpacking.

A similar form of camping, which isn't quite so concerned with weight but requires compact equipment, is motorcycle camping. Although motorcycles still suffer from the Hell's Angels image, it isn't really fair because the vast majority of motorcycle owners are dedicated family people who are very environmentally aware. The typical motorcycle club outing is an example of quiet, considerate campers who keep the campgrounds cleaner than the average car or RV camper, and you won't hear their motors roaring after dark.

Car camping: Car camping is the most popular form of camping because you carry everything you need in the family car and camp within sight of the car. This gives you more versatility than camping in any other vehicle because you can camp in the most elaborate campground with all the modern conveniences, or find an isolated spot along a country lane or toward the end of a logging road.

The bulk of information in this book is aimed at this type of camping, with some variations. Most car camping is done in designated campgrounds, ranging from national parks and national forests to other governmental or privately owned and supervised campgrounds. Many of these have resident rangers or managers.

Assuming that many readers will want to work their way into the more rugged outdoor sports, such as backpacking, we have placed a great deal of emphasis on lightweight and compact equipment. Even if you don't care to try the Appalachian or Pacific Crest trails, this equipment will serve you well for car camping because it occupies little space and won't weigh down the family car.

Even if you are traveling in a large RV, you will still appreciate sleeping bags that compress easily into stuff bags usual to find RV owners who are also backpackers, river runners, and die-hard conservationists. One of the major complaints against RVs is the amount of fuel they consume, but the gasoline mileage for some of the major brands has been gradually improved until now some

or may be left flat on the bunks that are folded away during the day. The same storage concerns apply at home, especially when equipment must be stored for months at a time.

Recreational vehicle camping: Abbreviated to RVs by nearly everyone, recreational vehicles can cost as much as a modest family home, or you can spend less than $1,000 on one. Along with the new lightweight and sturdy materials developed since World War II, RVs are an integral part of the camping boom in recent years. There was a time when RV owners and backpackers were no more friendly then Yankees and Rebels in 1865, but the lines between these two forms of camping are becoming more and more blurred, and it isn't un- models get better mileage than many family sedans.

TYPES OF RVS

Motorhomes: They are often called "land yachts" or "land cruisers," because they are at the top of the RV line, both in cost and in accessories. To most RVers, they represent the ultimate vehicle. Virtually every convenience of home can be included in a motorhome—all the kitchen gadgets, easy chairs, sound centers, queen-sized beds—and at night they can be divided into as many as three compartments for sleeping.

They can range up to 35 feet or more, although many are a stubby 18 feet. They are a great answer if you prefer the space of a trailer but don't like to tow vehicles. Also, they are much easier to back up and park than a trailer. They can be driven on all types of roads, and most have enough engine muscle to manage winding mountain roads.

While the majority of motorhomes are built on chassis from the top three or four truck manufacturers, a few are built from the ground up.

There are a few basic similarities, such as in overall construction. Most use a steel cage to frame and strengthen the walls and roof. You have a choice of steel, aluminum, or fiberglass on the outer walls, and you should have some kind of insulation between the outer skin and the inner walls of paneling for temperature control noise-deadening value.

The floor plans are different not only among various manufacturers, but also under the same brand. For sleeping privacy, most are divided by sliding doors into at least two compartments, and some offer three. Usually, but not always, there is a large sleeping area in the rear that doubles during the day as a set of facing gaucho seats with a removable table between. Others sacrifice storage space overhead for two rows of bunks across the rear where four children can sleep in comfort.

Most models have a couch or cafeteria-style table in the front compartment that folds out into a twin bed at night, and another bed can be hung from the top of the driver's compartment and clamped up against the ceiling and out of the way during the day.

There are rear-bath models and side-bath models, each with an equal number of advantages and disadvantages, depending on one's personal preferences.

Some have the kitchen area flush against the wall while more expensive models will have an L-shaped kitchen to separate it and to cut down on traffic through the food preparation area.

About the only place with no variation is in the driver's compartment, where plush, adjustable seats with all the instruments in a convenient location are more or less standard. One popular option is a double-sized chair on the passenger side where an adult and child can sit comfortably. Privacy curtains separate the driver's area, creating an extra sleeping compartment, plus shielding the driver from inside lights when on the road at night.

All motorhomes have at least one full-length closet, and storage bins are found throughout the vehicle, often along the sides on the outside, for skis or items that won't be used often such as spare parts and tools.

A motorhome's size can be a disadvantage when you decide to go exploring after finding a camping site. Many owners get around this by hauling along bicycles or trail bikes or towing a small vehicle behind. The main thing to remember here is that small cars with automatic transmissions should not be towed with the power wheels turning; a small trailer arrangement that holds the power wheels off the pavement is needed. Towing stick-shift cars in neutral causes no problems.

Pickup campers: Pickup campers have been a mainstay of the RV market for years, so much so that it is almost taken for granted that when someone buys a pickup, he or she eventually will buy some kind of "lid" for it, whether it be a complete camper with kitchen, bathroom, and beds, or a simple box that doubles as a cover for hauling building materials and as a basic shelter for weekend trips.

Four-wheel drive vehicles are increasing in popularity to explore the back country, so much so that many four-wheels come equipped with this basic camper already built on them. Also, a number of campers are being built specifically for the small imported pickups as well as the tiny ones made by domestic manufacturers.

Most campers fit right into the bed of a pickup and can be removed quite easily. Many owners have special racks built in their garages so they can back the truck in, remove the camper's bolts or clamps, raise it slightly, and drive it out.

There is a very fine line between a "lid" or a cover as the industry calls it, and the regular camper, and it probably isn't

slide in and are tied down, often project over the cab of the pickup, and this projection holds a double bed. The camper will have a galley-sized kitchen with a two- or three-burner LP gas stove, usually a cafeteria-style table that folds into another bed, plus a toilet and shower. The latter utilities are usually found only in the longer campers, the ones that project backward from the truck bed.

These models are a good compromise between a trailer and a mini-motorhome. They give you everything mini-motorhomes have (except a passage directly from the driving compartment to the camper) plus the option of taking the camper off the pickup and using the

all that important. Many do-it-yourselfers buy a bare shell and add their own equipment as they build in bunks, sinks, stoves, and lighting. Most covers, however, are no higher than the roof of the pickup cab (in order to reduce drag) which makes them too small to stand in, but just right for sitting, eating, and sleeping.

The more elaborate campers, which

truck for other purposes. Some of the largest campers offer an amount of space equal to that of a mini-motorhome and more space than in the increasingly popular van conversions.

Another benefit to a slide-in camper is that, should the occasion call for it, it can be jacked up at a campsite and left while you drive elsewhere.

Since both the increased weight and wind resistance increase your fuel consumption, many manufacturers have found a market for models that expand, or telescope, up and down. They are raised and lowered by either spring-arm lifter systems, hydraulic jacks, or electric motors, then lowered to rooftop level when under way. The advantage of this system is fuel economy; the disadvantage is that you sacrifice some storage and over-the-cab sleeping space.

The average slide-in camper is built on the same principle as the travel trailer-wood stud frames, and skins of either plastic or aluminum. Interior walls are usually wood or plastic paneling, with fiberglass or plastic foam for insulation between the walls.

Before you buy a camper or cover, be sure your pickup is built for the increased load. Dealers have "camper special" pickups for sale that have the stronger suspension, heavy-duty cooling system, bigger wheels and tires, and rack for the auxiliary battery that you need.

Mini-motorhomes: This type of RV is a compromise between the most expensive and the least expensive of the RVs. Until it was developed and made popular, there was quite a gap between the fleet-sized—and expensive—motorhome and the pickup camper. A mini-motorhome costs a bit more than a pickup camper, but less than a motorhome.

The primary difference between a mini and a regular, or standard, motorhome is the kind of chassis each is built on. A mini is built on a van or pickup chassis, and a motorhome is built on a modified truck chassis. This brings the price up considerably before anything is installed.

One major advantage over a pickup camper is that in a mini you can walk directly back from the driving compartment. Another advantage over other RVs is that the front doors are retained.

Minis range from 17 to 25 feet, sometimes more, and they are built lower to the ground than standard motorhomes. They feel much like a car feels from the vantage point of the driver's seat. They offer automatic transmission, power steering, cruise control, power brakes, and virtually every other option available in full-sized motorhomes.

Minis are approximately the same width as conventional motorhomes and travel trailers, about eight feet wide, and the gross vehicle weight (GVW) ranges up to 11,500 pounds. Those with the biggest capacities have dual rear wheels.

More and more manufacturers are using steel framing to enforce walls of fiberglass or aluminum, so construction is similar to that of full-sized motorhomes.

The interiors are often elegant, and always functional. You have a wide variety of floor plans to choose from, and each manufacturer often has five or six basic plans. In all models, the overhang section above the cab is a bed, but after that, you have a choice. Up to eight persons can sleep in a mini-motorhome.

The galley, bathroom, and living areas are as roomy as most motorhomes, and you have a choice of appliance sizes. All have a combination 12-volt to 115-volt electrical system with a heavy-duty alternator and a deep-cycle battery. If you want to install a power plant (generator) on your mini, it is possible to put one on the rear bumper, either as a removable or a permanent fixture.

It has either a demand or a pressurized water system that can be hooked to a city water supply. It has the water, sewage, and wastewater holding tanks, and everything else of this nature available on the big-brother motorhomes.

Minis give you superior gas mileage compared with motorhomes, and a definite break in price. On the other hand, you may find the ride a bit choppy since you're sitting directly over the wheels on a shorter wheelbase. You don't get the storage space of a motorhome, and you may find that luxury items, such as a blender and a microwave oven, are options rather than standard equipment.

Travel trailers: Year after year, this type of RV outsells all others and has formed the basis of the entire RV industry. In fact, the industry started with travel trailers soon after World War II, and their popularity hasn't diminished in the face of competition from other types.

The major reason for their popularity is that they can be towed behind your car or pickup to a campsite, then left while you explore the area or run errands.

You will probably find more variety within this type than in any other you investigate—from the simple, straightforward kitchen-and-bed models to the downright elegant ones that almost resemble a resort condominium inside. Some, such as the fifth-wheel trailers, even have spiral staircases.

Travel trailers come in various lengths, from the stubby twelve-footers that look almost as tall as they are long to the "park" models which go up to forty feet. The latter are not designed for frequent towing, however, as are their shorter relatives.

Each model has its own specifications for a towing vehicle, and you can match a small car to one as easily as a four-wheel-drive vehicle. However, you must make certain your tow vehicle is within the required limits, not only for safety but also to prevent burning out your transmission or engine.

Weight is the major difference between the types. The most expensive tend to be the lightest—those made of aircraft-aluminum skin. They usually hold together better over a longer period of time, while those made of wood-frame construction have a tendency to work screws and glue loose.

Camping trailers: Camping trailers are the most economical type of RVs, in terms of both initial cost and operating expenses. They can be towed behind nearly every type of car or truck because of their light weight and the fact that their low profile while under way creates virtually no wind resistance.

As most models have canvas sides, camping trailers sometimes are referred to as "tent trailers," which isn't a bad defi-

You will find that travel trailers offer every luxury of the fleet-sized motorhomes. The only disadvantage, depending on which state you're traveling in, is that passengers often are unable to ride in the trailers while they are under way.

Other than that, the sky and the thickness of your bank account are the limits on amenities. You will get a bathroom, kitchen, sleeping quarters, and both water and waste disposal systems. You will have a refrigerator and LP gas range, cupboards, and generous storage space. Options will include TV set, microwave range, air-conditioning unit and, of course, generator.

nition at all. Their simplicity gives them more of a camping-out feeling than any other type of RV.

This doesn't mean they are simply tents on wheels. Most models have a two- or three-burner gas stove, a sink with a water supply and city water hookup, cafeteria-style table, an icebox or small refrigerator, and lots of storage space. Some of the luxury models even have portable toilets and shower stalls! Some can sleep up to eight persons using fold-out wing beds at each end plus the convertible table and seats. Some come equipped with batteries to power 12-volt lighting, and nearly all have provisions for 110-volt hookups at modern campgrounds.

Nearly all camping trailers will fold down for travel lower than the tow vehicle, so low in fact that you may not need side mirrors for rear visibility.

The trailers are raised and lowered by two basic methods: one system uses a spring-arm mechanism which you guide up and down, and the other is a crank system—the more popular of the two.

Don't be fooled by their apparent small size when you see one being towed down the interstate or stored away in a friend's garage. When they are erected and the bed wings folded out (some even fold out on the sides, too), you will see that their size almost doubles. There is plenty of headroom for average-sized people, and a family of four can easily sit inside one while meals are being cooked—or just to get out of heavy weather.

Some families just getting involved in the RV way of life buy a camping trailer first, then work their way up through the various types of RVs. Others, however, wouldn't consider owning any other kind of RV and point to the advantages of storage, ease of towing, and a low inital cost which ranges from less than $1,000 up to around $5,000.

TRAILER CLASSES

Manufacturers have placed trailers in four classes by weight. The buyer is able to determine from these classes what kind of tow vehicle is needed for each type of trailer.

Class I—Light duty includes the tent trailers as well as small trailers for snowmobiles and trail bikes. The gross trailer weight is under 2,000 pounds, and the tongue weight is up to 200 pounds.

Class II—Medium duty trailers are usually those with single axles in the small- to medium-size range. Gross trailer weight is 2,000 to 3,500 pounds, and the tongue weight is 10 to 15 percent of the gross trailer weight.

Class III—Heavy duty includes the dual-axle models, which have a gross weight of 3,500 to 6,000 pounds and a tongue weight of 10 to 15 percent of the gross trailer weight.

Class IV—Extra heavy duty includes the fifth-wheel models and those with a gross trailer weight of 6,000 or more pounds and 10 to 15 percent of the trailer weight on the tongue. Nearly all trailers in this class require a pickup for towing, and of course all fifth-wheel trailers must have a pickup for towing.

THE RIGHT HITCH

Bumper-mounted hitches are not recommended for towing any type of trailer. Always use either hitches mounted to the underbody of your tow vehicle or, in the case of heavy loads, hitches mounted directly to the frame.

For trailers weighing more than a ton, load-equalizing hitches should be used. These are designed to distribute the weight between the axles of both the trailer and tow vehicle.

For pulling the light Class I trailers, use a weight-carrying ball hitch that takes the trailer's full tongue weight. These are not to be mounted on the bumper, though, but always on the frame or underbody of the tow vehicle.

Safety chains should always be installed. The chains should cross under the trailer tongue to prevent it from dropping to the road in case of hitch failure. A breakaway switch should be used for a trailer that has brakes. The breakaway switch automatically applies the trailer's brakes if the hitch fails. These switches are available in both electrical and mechanical models.

Towing tips: Keep the trailer's center of gravity low for good stability. It's the same as ballast in a boat.

Stow heavy articles such as canned goods, tools, and books as near to the floor as possible. Stow lightweight items such as clothes and linens on a higher level. Never store dangerous items overhead.

Distribute the weight of the load evenly to keep the trailer from tilting or leaning and to keep from putting too much strain on the hitch and the towing vehicle. Balance the load from side to side, too.

Carry only essential items and keep the load down. Be sure all doors and drawers are secured. Leave nothing loose.

Van conversions: Vans have become the "in" vehicle during the past two or three years, and owners, particularly young people, often turn them into pieces of moving artwork. You see them daily with imaginative paintings on the sides and big stretches of glass, often one-way or privacy glass. Look inside them and you'll see furnishings fit for a king: deep shag carpeting, plush seats, small stoves, refrigerators, TV sets, and stereos.

Of course, not all converted vans are so richly decorated. Most are rather basic with everything other RVs offer—only in less space. However, vans have an advantage over most other RVs: they can be used daily for going to work, running errands, and so forth. Some call them "super stations," because their size isn't a problem when you are looking for a parallel parking space, yet they are large enough for RV vacations.

Since there isn't enough headroom for comfortable conditions while using them as an RV, many owners put a higher roof first on the priority list. There are two choices: a permanent extender can be put on to give you headroom, or you can have an expandable roof that is raised and lowered while you're camping. The permanently expanded roof has the disadvantage of creating more wind drag, thus decreasing gas mileage, but it gives you permanent storage space and doesn't require passengers to crawl about while under way. The expandable roof creates virtually no additional drag and doesn't make the van a broader target for side winds.

Fifth-wheel trailers: The newest member of the RV caravan is the fifth-wheel trailer, first introduced on the market in the early 1970s. It was an instant success, and hardly an overnight campground is without one in residence.

Essentially, fifth-wheelers increase the living quarters by about seven feet without increasing the wheelbase. And they give you a split-level home.

The major differences between them and standard travel trailers are first, the type of hitch required, and second, the extra level above the hitch that is the bedroom. The extra seven feet are above the pickup or four-wheeler that is used as the tow vehicle. Thus a 29-footer has virtually 22 feet of space behind the tow vehicle.

The hitch is designed to be anchored over or just in front of the rear axle of the pickup. It sits in the bed of the pickup and is attached to the frame for absolute security. If you've noticed a semi trailer turning, you'll get an idea of how short a turn your fifth-wheeler can make, because the pickup can pivot at a right angle from the trailer.

Fifth-wheelers also are less prone to swaying and jackknifing because of the stability of the hitch arrangement and in part because of the weight taken by the pickup.

Hitching up is easy. The trailer has a kingpin that hangs down and couples to the hitch in the pickup bed. You have excellent visibility through the pickup window since the hitch is almost at eye level—unlike standard travel trailers that almost always require a second person standing outside giving you backing instructions. And manufacturers, pointing to the safety record of similar hitches on interstate trucks, claim this kind of hitch is the safest.

Fifth-wheelers are usually built of the same type of material as the standard travel trailers. They should be built sturdily, as their increased size makes them more likely to be used as permanent residences by owners. In addition to the living space available, many manufacturers have added such amenities as sliding glass patio doors and glass alcoves that crank outward to form bay windows, with seats, so guests can sit without having to move their feet every time someone walks the length of the trailer.

Parked, and with everything set for a long stay, there is little difference between a fifth-wheeler and the most luxurious permanent mobile home. And you have the advantage of being totally mobile. Crank in the alcove, hitch up to the pickup, and you're off.

Fifth-wheelers will sleep up to eight in comfort. But many of these RVs are owned by couples on the move, who opt for sleeping accommodations for four and use the rest of the space for their other needs, such as a bigger kitchen or a larger living room.

The typical fifth-wheeler has tandem axles for extra support and stability while traveling, your choice of a double or twin beds in the master bedroom upstairs, a bathroom with both tub and shower, a totally equipped kitchen with double sink and options including a blender and a microwave oven, a TV cabinet and antenna option, gauchos that fold out into another double bed, a dressing room just off the master suite, and lots of storage space for both food and clothing.

How much current is drawn? When you consider size in the auxiliary batteries and generators you need to buy for your RV, it is best to first understand the amount of current each electrical fixture in your RV draws. Usually the batteries and generators that come as standard equipment are sufficient for normal use, but many RV owners use more than the normal load. A large group, a pack of small children, or unusual circumstances can cause the system to be overloaded. Your battery might go dead in the middle of the night, cutting off your furnace, or the generator might rebel and refuse to operate until you cut back on the demands made upon it.

Following is a list of wattage draws on the 12-volt system:

- Dinette light — 3
- Hall light — 3
- Reading light — 1½
- Kitchen light — 1½
- Each compartment light — ½
- Exterior light — 1½
- Step light — ½
- Bathroom light — 6
- Range hood and vent fan — 5½
- Furnace blower — 6
- Water pump — 6
- Monitor panel — 2½
- Vent fans — 3

Wattage for 110-volt equipment:

- Air conditioner — 1,100–2,200
- Electric heater — 1,100–1,540

CAMPER'S SECRET

If you plan to visit at least half a dozen federal camping areas in one year, particularly national parks, the Golden Eagle Passport is a good buy for you and your family. It costs ten dollars a year and covers entrance fees for everyone in a single vehicle. It's not good for camping fees, however.

The Golden Age Passport picks up where the Golden Eagle leaves off. This is for persons sixty-two and older to replace the Golden Eagle. It has an added advantage as it also gives you a 50 percent discount on overnight camping in all federal camping areas.

Both passes are available at National Park and National Forest Service offices throughout the country.

ESPECIALLY FOR FIRST-TIMERS

- Stereo 220
- TV 330
- Microwave oven 1,430

When choosing a generator for your RV, keep these figures in mind to be sure the generator is strong enough for the load. They are rated in the amount of watts produced:

- 2,500—Powers one air conditioner and all lighting
- 4,000—Powers one air conditioner, all lighting and one appliance such as a toaster or electric frying pan, or two air conditioners and nothing else
- 5,000—Powers two air conditioners, all lighting, and one appliance
- 6,500-7,500—Powers two air conditioners, all lighting, and two appliances

Weighing in: It is very important that you keep your RV loaded within the requirements established by the manufacturer. In overloaded RVs fuel consumption increases, and structural damage can occur. The RV becomes a safety hazard as well.

Of course the simplest way to determine weight is to go to a trucking company and have your RV weighed. But if that is inconvenient, you can figure your weight, using the following list of sample weights as a guide:

- Tableware for eight 5 pounds
- Electric coffee pot 4
- Portable toilet 18
- Set of tools 10
- 12-inch TV 20
- Toaster 4
- Sleeping bag (not down) 5
- Standard tent 50-60
- Portable lantern 6
- Air mattress 6

Most states do not require you to stop in at highway weigh stations.

Sample weights: Here are some sample weights of RV fixtures (and people) to help you estimate the *gross vehicle weight*, which is the RV's weight plus that of all it is carrying. Your manufacturer will provide you with the optimum GVW figure for your particular RV.

Four passengers (two adults, two children)—500 pounds
Clothing for four—100 pounds
Food for normal circumstances—200 pounds
Auxiliary (20-gallon) water tank (filled to capacity)—200 pounds
Extra (20-gallon) fuel tank (filled to capacity)—160 pounds
Auxiliary battery—49 to 140 pounds, depending on amperage
Auxiliary deep-cycle battery—56 pounds
Towing hitch—40 to 125 pounds, depending on class of trailer
Trail bike—150 to 250 pounds
Bunk ladder—6 pounds

The energy problem: You've undoubtedly heard the complaints about how RVs are fuel hogs, and there is no escaping the fact that your car uses more fuel towing a trailer, and a motorhome uses more fuel than does a car.

What you've heard isn't the whole story. The Recreational Vehicle Industry Association (RVIA) has shown that your family on vacation in an RV actually consumes less energy than it would at home. In the first place, your family won't be using the electrical energy at home, nor will you be utilizing the water and sewage system. By comparing your gasoline consumption during the same length of time driving to and from work, running errands, and so forth, you will be surprised how little difference there is between what you consume on vacation and what you consume under everyday conditions.

Another factor to consider is that many family vacations in RVs are taken reasonably close to home, and the RV is often used in lieu of a motel or resort room. Many vacations consist of driving to one spot and staying there for a week or more, then returning home.

Forget about the energy problem a moment and concentrate on costs in general. Statistics from the RV industry show that the typical family of four will save more than $1,000 a year on travel by using an RV. Lodging alone is a big factor in saving: most commercial campgrounds charge no more than $7.50 a night. Compare this with $30 a night for motel charges during the tourist season —often more—and your vacation is already considerably less expensive.

Most families using an RV on vacation prepare for the majority of their meals in the RV, treating themselves to a breakfast and an occasional dinner out. It is simply the difference between eating at home and eating in a restaurant. Many families enjoy getting up early and preparing a pot of coffee and some orange juice, driving an hour or two, then having brunch in a restaurant. Such treats are part of the joy of traveling; you can sip your coffee and juice while on the road.

With all vacation costs taken into consideration, it is a rare family that can't take a trip in an RV and save money, no matter how far they travel. All reputable commercial campgrounds have free recreational facilities such as swimming pools, horseshoe pits, tennis courts, and playground equipment. Often the campgrounds are near skiing, either downhill or cross-country, for winter guests.

Other RVers find places away from established campgrounds to park and unload their boats and water skis or fishing equipment. They can have a family outing on a lake or river that ordinarily would be possible only if they owned a summer cabin.

SANITARY SYSTEMS

There are five major types:

Permanent freshwater flush toilet—This is the most common and the easiest to install and use. It is virtually odor free and uses no electricity. On the minus side, it requires a holding tank with chemicals to be added, an extra supply of water for flushing, and can freeze in skiing weather.

Electric recirculating—This operates on a small amount of recirculated water, needs emptying only every three to five days, and doesn't require a special holding tank. However, it requires 12-volt power, is one of the most expensive types, requires special chemicals after each dumping, and often gives off a strong chemical odor.

Incineration—This is often an option for the more expensive RVs. It doesn't require a holding tank, uses no water for flushing, uses no chemicals, won't freeze in the winter, and can handle sanitary napkins and disposable diapers. But if the battery is low it won't function properly, and it needs at least fifteen minutes between uses to operate properly—a problem with large groups and at certain times of the day when use is heavy.

Waste destruction system—In this type, waste is burned during travel or when the main engine is running while standing. When used constantly and properly, the holding tank won't have to be emptied. And since the waste is burned, there is no danger of bacteria causing contamination. The disadvantages include the need for a holding tank, the need of liquids to hide odors, and the fact that if the system cools down while in use, it won't work.

Portable toilet—This is the most basic type of toilet. It is the least expensive and can be used not only in RVs but in cottages, at home, or in the sickroom. Many people feel this toilet far too primitive, however, since the waste must be carried to a dumping station and the whole system cleaned before reuse.

PLAYING IT SAFE

Since an RV is a combination home and vehicle, the potential for accidents and injuries definitely exists. But by following basic safety rules, you can make your vacations and weekend trips as much fun as you hoped them to be.

The first safety tip comes from the RV industry, and that is to be certain the RV you buy or rent has the Recreational Vehicle Industry Association (RVIA) standards seal. This seal informs you that the RV has been built according to the basic standards of the industry as a whole.

While safe operation of vehicles rests with the driver, there are basic rules for manufacturers to follow which are set down by the American National Standards Institute. The standard for RVs, called Standard A119.2, contains more than 500 requirements for plumbing, heating, and electrical systems.

Here is a brief rundown of the standards:

Fire and life safety—The vehicle must come under the minimum flame-spread ratings for interior walls and ceilings. A vehicle with fuel-burning appliances or an internal combustion engine must have a fire extinguisher with the proper charge. Also regulated is the location of the gasoline filler spout so it is away from potential ignition sources such as the exhaust pipe.

Plumbing systems—This standard prevents the use of substandard materials that would cause contaminated water. The drainage system could be hazardous if not installed properly, and the specifications prevent sewer gases from accumulating. Standard-sized pipes to insure adequate flow of water are required, along with sanitizing instructions, an adequate drainage system, adequate venting, and proper drain outlets. A final testing of the complete system is required.

LP gas systems—This standard involves safety-relief valves for excessive pressure, controls the location of the LP gas container, provides specifications for all pipe sizes, covered tubing, and appliances, and tests for leakage. Sufficient clearance is required to prevent ignition of adjacent surfaces.

Electrical systems—Essentially, the same requirements for an RV that apply to a home are used.

In addition to the RVIA standards, there are the normal common sense safety precautions one should take while traveling in an RV. These apply especially to the self-contained mini-motorhomes, fleet-sized vehicles, and van conversions.

Since most RVs aren't designed for

drag racing, you must always allow extra space between yourself and other vehicles. If you are following one, remember that your extra weight will lessen the vehicle's stopping power. It is not what you are accustomed to in the family car.

The same applies to passing. It will take you longer to get around that car ahead. You are driving a vehicle of greater length, so allow plenty of space before you pull in ahead. Learn to "read" your right-side rear-view mirror at a glance. And always watch that potential blind spot right beside you. Unless you have a convex mirror, or a small stick-on convex one attached to the bigger mirror, a car can pull up beside you and you'll never see it.

The first time you drive or pull an RV, you will understandably be a bit nervous. Test it on familiar roads before taking off on the interstate and by all means practice backing up. In spite of what you may think, you will be very lucky indeed to find only campgrounds with pull through campsites. Sooner or later you will find yourself in a spot where you have to back up the RV, and nothing is more satisfying than being able to do so without bumping into a picnic table, knocking over the water pipe and starting a gusher, or having a crowd gather to watch the fun.

WHICH IS BEST FOR YOU?

Generally the type of camping you set out to do will depend on the equipment you own. But if you are thinking of renting or buying special camping vehicles, consider the points listed.

Recreational Vehicles

Advantages
All the comforts of home
Don't have to pack and unpack
Can lock up when going on a hike
No worries about space
Can cook anything you want

Disadvantages
Most expensive form of camping
Isolated from the outdoors
Limited to larger campgrounds
Don't learn about the outdoors
Can't pack and go in a few minutes

Car Camping

Advantages
Great mobility in choosing site
Gives family true outdoor experience
Teaches children self-sufficiency
Easier to find isolated camping areas

Disadvantages
Requires more advance planning
Requires more compact equipment
Requires more outdoor knowledge
Car is crowded en route

While we're on the subject, learn to back up using only the mirrors. It is easier once you learn, and the mark of a professional.

If your RV or tow vehicle is equipped with a speed control device, by all means use it. Most RV owners plan to drive a bit below the 55-mph speed limit (some states require it) for safety and consideration. And keep to the right. You present an imposing target if you're driving in the fast lane and those following you have a restricted view ahead.

If you are driving on a two-lane mountainous road with frequent turnouts, keep an eye on the traffic piling up behind you and turn out to let it pass whenever possible. Many states require you to use a turnout if five or more vehicles are following you.

You should always compensate for sudden wind currents, such as those encountered when passing a large truck or emerging from a tunnel or a viaduct. It is best to have a firm grip on the wheel in these cases, and increase your speed slightly rather than following your first instinct of slowing. Treat wind gusts much the same as driving on ice—keep power on.

A WORD ABOUT PETS

Few subjects get campers' backs up more quickly than pets, yet the world seems divided almost evenly between those who wouldn't leave home without the family pet and those who wouldn't think of taking a pet camping. It is clearly a situation to which there is no solution that will satisfy every camper. The National Park Service has for a number of years banned pets from its campgrounds and trails, but they are welcome at nearly all other campgrounds.

The main issue, obviously, is the same with all outdoor etiquette; your neighbors should hardly be aware you are present, which is impossible if your dog is a barker or loves to cruise the garbage cans. The general rule of thumb is that if you've had complaints about your dog at home, then leave it there when going camping.

On the other hand, there is no more comfortable companion to have on a camping trip than a well-trained and lively dog. Parents rest easier when their youngsters leave their sight with the dog accompanying them. Dogs can keep many of the camping pests, such as food-stealing ground squirrels, away from camp and serve as the first line of defense against other creatures. Just hope, of course, that your dog knows enough about skunks and porcupines to bark at them from a safe distance.

Many outdoors people travel constantly with their dogs and make or buy packs for the animals so they can at least carry their own food on outings. A child who decides all adults and siblings are wretched people will work his or her way back into the flow of things if a dog is around for affection and as a companion with whom to file complaints.

The basic rules of most neighborhoods should apply to camping with a dog: it should be confined to the family's camping area, on a leash, and should not be permitted to roam through the entire campground, setting pet haters' teeth on edge.

Cats present another kind of problem for camping trips. Although they seldom cause many problems in the campground, they do have a tendency to wander off and get lost. City cats are notorious for having absolutely no sense of direction. Sometimes, out of fear of a dog or from sheer exuberance, they decide to climb a tree. Then they get stranded where nobody can retrieve them. Their independent nature and lack of innate wilderness know-how can make them more of a problem for owners than their dogs.

Since it is easier to leave cats for several days alone in a house with an ample food supply and a litter box handy, it is best to leave them home. If they could talk, they'd probably thank you for it.

PREPARING FOR THE NEW EXPERIENCE

It is very important that you and your family think of camping as perhaps the most natural form of recreation available. It is not a venture into the great unknown; on the contrary, it is simply doing what people have always done and always will. Millions throughout the world live all their lives doing what we call camping, many by choice. Camping is as natural as a backyard barbecue; only the location is different.

Granted, some people approach camping with vague fears, especially children who are reared in urban areas and know nothing more about the outdoor experience than what they learn in the local playgrounds surrounded by streets and buildings. In these cases, and with some apprehensive adults, you may have to ease into the sport gradually.

One method of overcoming this is to find a nearby campground and make one or more trips there for picnics and day hikes. Let the family become acquainted with the area and see how much campers are enjoying themselves.

CAMPER'S SECRET

Noise is one of the worst problems in busy campgrounds, and most supervised ones have hours posted when all radios, phonographs, and tape decks must be turned off. It is more of a problem in the less developed campgrounds, such as many in national forests. RVs with electrical generators that operate the air-conditioning system and other 110-volt appliances are one of the worst offenders. You should plan on turning it off no later than when darkness falls, and preferably more than an hour earlier since most campers associate the early evening hours with silence. If you're a tent camper near an RV with a generator, don't hesitate to ask the owners if they'd mind shutting it off. Most will comply.

Encourage them to talk to the campers and ask questions about their equipment, where they've camped, and what they do while camping. Stay until after dark so they can see the campers settling down for dinner and bed. This will remove much of the mystery of camping, and in many cases will make the trip home a bother because they will wish they were still there for the night. On these occasions it is easier to sell first-timers on the idea that food and drink taste better when brewed over a campfire or campstove, and that you sleep better outdoors.

In other words, you often have to sell camping to first-timers because it is in our nature to approach new experiences with misgivings. Children tend to be very negative about a new experience, whether it's an unfamiliar dish, a first meeting with a cousin, or trying a new form of recreation.

When planning a camping expedition, involve **everyone** totally in the trip preparations. Take the whole family to the outdoor store to look at equipment. Let them read this book and others on the subject. Involve everyone in the menu planning (but with the understanding that parents have the last word on nutrition, of course). If it isn't a total family effort, someone is likely to feel he or she is being dragged along.

Be democratic, too, once on the road. While parents may want to find a campsite tucked away from the crowds, children usually like to be around other children. Thus, parents may have to forsake total peace and quiet and isolation in favor of happy camping children. Many of the larger state parks and national forest campgrounds have common playground areas where children can strike up new friendships while playing team sports. If you choose your campground carefully, you will be able to combine your desire for privacy with the children's desire for action, since most large parks offer large campsites. Check first with the ranger or friends and see if campsites are crammed together or if they have a shield of plants or other obstructions between them.

For specific information about choosing a site, see PART IV, RESOURCES.

OUTDOOR GOOD MANNERS

It is difficult to improve upon the Boy Scouts of America's Outdoor Code, which follows:

"As an American I will do my best to:

"Be clean in my outdoor habits. I will treat the outdoors as a heritage to be improved for our greater enjoyment. I will keep my trash and garbage out of America's waters, fields, woods, and roadways.

"Be careful with fire. I will prevent wild fire. I will build my fire in a safe place, and be sure it is dead out before I leave.

"Be considerate in the outdoors. I will treat public and private property with respect. I will remember that use of the outdoors is a privilege I can lose by abuse.

"Be conservation-minded. I will learn how to practice good conservation of soil, waters, forests, minerals, grasslands, and wildlife; and I will urge others to do the same. I will use sportsmanlike methods in all my outdoor activities."

Children learn outdoor good manners from the example of their parents. If cutting dead trees is banned in your camping area, don't make an "exception" because you love big fires. Keep your campsite clean and in better condition than you found it.

ESPECIALLY FOR FIRST-TIMERS 35

A few specific items might be added to this admirable code of outdoor ethics, including:

Radios: They're handy for keeping up with the news of the world (unless you're easily depressed) and local weather reports. But it is virtually impossible to hear the wind in the trees and the birds singing if you or one of your neighbors has a radio going full blast. One small transistor with a fresh battery used for information rather than entertainment is all you really need on a camping trip.

Lanterns: Part of the pleasure of camping is sitting on a log or stump, or the ground, watching day end and night begin, watching the stars appear one by one and listening to the night sounds that are quite different from those during daylight hours. Then when it is totally dark and your eyes have adjusted, you can go to sleep in the soft darkness. Thus, lanterns have a way of interfering with your enjoyment and should be used only when necessary. Too many of the gasoline lanterns make a constant hissing noise that carries long distances and interferes with your neighbors' enjoyment of the evening.

Power toys: This includes scooters, motorbikes, motorcycles, chainsaws, generators, and the like. Campers appear to be divided into two distinct camps: those who love power toys and those who hate them. One child on a motor scooter can make life miserable for an entire campground, and if you are a silence lover, you may have no other recourse than leaving in search of a campground where these toys aren't popular or where they are not permitted. Many campgrounds have a 10 P.M. curfew on noisemakers, and you should honor that rule. Better yet, set your own curfew before darkness falls.

Vandals: Always report vandalism to the campground ranger or manager. Strangely, some parents strongly resent other campers implying that their child (or spouse or even themselves) is committing an act of vandalism, so it is best not to take matters into your own hands and try to deal with it directly. Vandalism is one of the major problems in the outdoors, as is its cousin, littering.

PART II
GETTING READY

CHAPTER 3

CAMPING EQUIPMENT— THE ESSENTIALS

It wasn't too many years ago that a camping trip took on the characteristics of a trek across the Oregon Trail in a covered wagon. The pioneers in family camping as a form of recreation had to make do with many standard household items that were manufactured with no concern for weight or bulk. Cooking equipment included cast-iron skillets, steel pots and pans, and second-best tableware. Conspicuously absent were devices such as nesting pots and pans or tableware that collapsed into small components for ease of carrying. Stoves small enough to fit into a coat pocket were rare.

Shelter was often a heavy cotton or canvas tent that was too heavy to carry more than a few feet at a time and was subject to mildew and leaks late at night.

Clothing was bulky leather coats or heavy woolens, which got increasingly heavy the longer the rain fell. Rain gear was both stiff and heavy.

With today's ultralight insulated clothing and sleeping bags, and the new fabrics for garments and tents, these items occupy little space and weigh only a fraction of their earlier counterparts. Now it is possible for a family of four to go on an extended camping trip and carry everything they need for a week in a medium-sized car. Dehydrated and freeze-dried food has been improved to the point that a dinner for four takes no more space than a pair of rolled jeans and weighs less.

In order to keep your investments in equipment to a minimum, it is best to purchase the ultralight and most com-

40 FAMILY CAMPING

CAMPING EQUIPMENT—THE ESSENTIALS 41

pact equipment available. This will enable you to use the same equipment on all camping trips, whether you're in an RV or car camping, backpacking or bike touring.

Less is best. As an illustration of how revolutionary the development of down-filled products was, Eddie Bauer recently compared the weight of the outdoor gear he used prior to his invention of quilted down garments and sleeping bags in 1935.

good woolen clothing will last for years and years. Thus it is not uncommon to see dedicated campers wearing articles of clothing that have been washed and mended so many times that the clothing looks as though it were borrowed from a stage hobo.

Campers become loyal to their equipment, at times superstitiously so, and they do not worry about things looking new. Often a reverse snobbery results, and those campers who would

---

THEN
One pair of Hudson's Bay blankets, 72" × 90", pinned into a bag 36" × 84"; comfortable to 10° F. without campfire; subzero with fire: 12 lbs.

Alpaca lined hooded parka good to 10° over wool shirt and wool underwear: 4 lbs.

Malone mackinaw pants worn over wool underpants. Comfortable to 10° without fire: 4 lbs., 12 oz.

Total: 20 lbs., 12 oz.

NOW
Skyliner goose down-filled bag, 72" x 32". Comfortable to -30° F. without fire: 7 lbs., 12 oz.

Eddie Bauer Expedition Parka worn over medium-weight wool shirt and undershirt; comfortable to -40°: 2 lbs., 13 oz.

Cotton field pants worn over down underpants plus duafold two-layer underpants. Comfortable to -20°: 3 lbs.

Total: 13 lbs., 12 oz.

Not shown in this example is the equally major savings in bulk. All the down-filled items can be compressed into small stuff bags, saving at least one-fourth of the space taken by the heavier woolen items at the left.

Most investments in camping equipment are one-time expenditures if you shop carefully. With proper care, tents and cooking equipment will last for decades. Shoes, boots, socks, and some other articles of clothing will eventually wear out, of course, but footwear can be repaired repeatedly, and insulated and

never think of wearing a shirt or blouse with frayed cuffs and collars at home think nothing of wearing a tattered parka with ripstop nylon patches decorating it. These clothes become the on-trail equivalent of the now-classic tweed and corduroy coats with elbow or shoulder leather patches.

Owing to a combination of space-age technology and intense competition among manufacturers, plus rigid requirements by suppliers who deal directly with purchasers, outdoor gear is among the most carefully designed and

CAMPING EQUIPMENT—THE ESSENTIALS

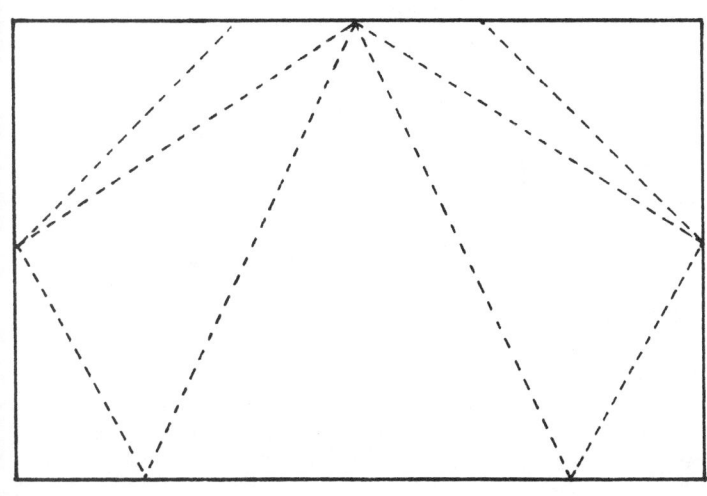

Eddie Bauer designed a simple tent made of a single piece of waterproof "balloon cloth" that could be used in a variety of shapes— half-pyramid, A-shape and lean-to. Such tents are still popular with many campers.

manufactured products on the market. It is built to withstand the widest extremes in temperature and moisture in the world, and the roughest use. Thus, when you buy a tent or a stove or a backpack from a reputable dealer, you can expect it to have a longer life than almost any other product you buy.

Basically, your needs for camping are centered on the essentials for sustaining life anywhere: food, clothing, and shelter. You can make these needs as basic or as complex as you wish, and with all the technology involved in manufacturing and processing these essentials, the choices become wider, making it difficult to draw up a list that will be satisfactory for every camper.

Following is a list of basic equipment needed for virtually every camp, whether it is a drive-in campground or one in the back country. The list is also suitable for any number of campers in a group, or even a solo camper.

A family of four or six can carry all the camping equipment and food in the family car, provided that equipment is made of ultralight, compact modern materials such as goose down and nylon.

EQUIPMENT CHECK LIST

The following check list of camping equipment is broken down into the needs of each individual camper and the equipment needed by the group as a whole. Some overlap is inevitable.

For Each Individual

- sleeping bag
- insulated ground pad
- plate
- knife, fork, spoon
- cup
- water bottle
- watch
- suntan or sun-screen lotion
- toothbrush, toothpaste
- comb
- soap
- mirror
- towel and washcloth
- insect repellent
- hooded parka or rainshell
- cap or hat
- sweater (light wool or goose down)
- long pants
   cotton for warm weather
   wool or wool blend for cool weather
- rain jacket and pants (or poncho)
- shirts
   cotton for warm weather
   wool or wool blend for cool weather
- sturdy shoes and boots
   boots for hiking
   lightweight shoes for camp
- socks (lightweight and wool)

CAMPER'S SECRET

Many professional guides wouldn't think of leaving on a trip without a pair of pliers hanging from their belt in a leather sheath. Pliers are indispensable for many camping chores. You can use them as pot holders, to move the grill around, to remove stubborn bottle caps, as tongs, and for a host of other cooking tasks. They can be used for repairing clothing and tents. They can double as a light hammer—for almost any task that calls for a wrench or hammer. The longer you carry a pair, the more uses you will find for this simple tool.

CAMPING EQUIPMENT—THE ESSENTIALS

- underclothing
- gloves or mittens

For the Group
- tent(s)
- stove and fuel
- nesting cook set
- pot holder
- tableware
- can opener
- cook spoon
- food and seasonings
- salt, pepper, sugar
- beverage mixes
- spatula
- grate
- water bucket
- dish scrubber
- tarp or dining fly
- toilet paper
- first-aid kit
- flashlight with spare batteries and bulb
- camp lantern
- reading material, games, etc.
- backpack
- repair kit (needle, thread, etc.)
- rope (for guy lines, clothesline, etc.)
- small hand-operated winch
- lashing cord for packing
- automobile repair kit
- emergency provisions (see "The Ten Essentials" below)

Optional
- camera and film
- a few coins for phone calls
- protective lip cream or stick
- pajamas
- inflatable pillow

The Ten Essentials
- extra food
- extra clothing
- matches in waterproof container
- candles or chemical fire-starter
- first-aid kit
- maps in waterproof case
- compass
- flashlight with spare batteries and bulb
- pocketknife with sharpening stone
- sunglasses

The importance of these Ten Essentials cannot be overemphasized. You will find yourself needing them while enjoying the outdoors in all kinds of weather conditions, from the best to the worst. So make them as much a part of your camping experience as coffee or hot cocoa, whether you are camping in a crowd at a private or public campground or miles away from the nearest human being.

Extra food: You should always keep an extra day's supply of food for each person in your party that is not to be touched except in emergency situations. Some campers have been known to select something particularly untasty to be certain their will power isn't put to an impossible test. Obviously, candy bars, trail gorp, or other tempting items should not be in the emergency provisions, but lightweight items such as freeze-dried or dehydrated food should be in the package.

Extra clothing: You should always have dry and warm clothing to wear in case a sudden storm blows in or you take

---

CAMPER'S SECRET

Most campers always take more gear than they will need. One way to eliminate this unnecessary gear is to sort through all the equipment after the trip and see what you didn't need (the Ten Essentials excepted).

an unplanned dunking while fording a stream. Always keep warm, dry clothing in your pack, even if it means wearing dirty clothing an extra day or two. Be sure it is wrapped in waterproof plastic inside the pack.

Matches in waterproof container: Small waterproof containers for matches should be kept filled and tucked away and never used except in emergencies. The best match containers are made of steel with a rough striking surface on the sides or inside the lid. Again, do not use except in emergencies.

Candles or chemical fire-starter: Fire-starting items are almost as important as extra food and clothing. A few stubs of household candles scattered throughout the camping equipment, or one or two full-sized candles, are always welcome for starting fires with wet wood. Better yet, buy some chemical fire-starting pellets or strips of chemical fire-starter, and never use them for ordinary fire-starting. Neither type takes much room in a pack.

First-aid kit: This should include adhesive bandage strips for small cuts, gauze pads of various sizes, a full roll of adhesive tape, salt tablets, an analgesic, a needle for treating blisters, and a first-aid manual. It is also recommended that you carry a sheet of moleskin for covering blisters, a single-edge razor blade, water purification tablets, first-aid cream, a mild antiseptic such as Bactine, antihistamine tablets, and a snakebite kit. (See CHAPTER 10 for a complete first-aid check list.) Of course, you should carry any prescription drugs you must take regularly or in case of emergencies.

Maps: These include road maps and topographical maps of the offroad areas you are visiting. Most recreational stores stock the "topo" maps or their equivalent, such as national forest or national park maps. Store them in waterproof cases, such as clear plastic cases with zippers or snaps. Also, be certain you know how to read them. Instructions are in CHAPTER 9.

Compass: As with maps, a compass is of little value unless you know how to use it in your area and with the maps you buy. Learning this can be a form of camping activity for the whole family. (See CHAPTER 9.)

Flashlight with spare batteries and bulb: Few emergency items are more likely to be rendered useless by campers than flashlights. Some campers find it impossible to face bedtime without reading a chapter or two of a book. Always keep one set of batteries and a bulb sealed and tucked away.

Knife and sharpening tool: Some campers get carried away with knives and wear a bayonet-style belt knife or an enormous Swiss army knife with a score or more blades. The best belt knives have short blades, and the best pocketknives are based on the old Boy Scout knives with a big blade, a can opener, an awl, and a screwdriver. Some of them have a blade that doubles as a can opener and screwdriver. There's nothing wrong with the biggest Swiss army knives, but for emergency purposes, the smaller pocket knives will suffice. A small sharpening stone should be carried.

Sunglasses: The only camping site that might not require sunglasses as an emergency provision is a deep, dark forest. However, they should always be carried. In the case of camping in open country, in the desert, and in snow, they are absolutely essential. If you wear prescription glasses or contact lenses, be certain your sunglasses are prescription also. Keep with your glasses an elastic holder to keep them from falling off.

EDDIE BAUER'S CHECK LIST

Now compare this list with one compiled several years ago by Eddie Bauer when he was outfitting hunters, fishermen, mountaineering expeditions, polar explorers, and many others.

"Because thousands of young outdoor people have asked what I have personally found necessary on my treks into remote areas, I have prepared a check list of those items. Some were homemade, but few people will go to the trouble of making their own today. During my years of outfitting I have made it my business to field-test all the new products, including many that I have created, to meet the needs of our constantly changing public taste, environment, economy, and other factors.

"It's a special joy to help those who choose to rough it, to leave the beaten trails, to hold costs down, to live off the land as their forebears did.

"Essential to me during my long life of outdoor outfitting and outdoor living are certain items I depended on in the backwoods of California, the Rockies, the Pacific Northwest, Canada, and Alaska. These items need not be costly, just simply adequate whether going solo or with a companion. As a check list, it has been useful to me, my friends, and customers. It is wise to anticipate one's needs for an outing, and this list may prove useful in selecting certain items that are often overlooked or forgotten.

"I suggest you start a list of your own. My list started in 1920, and it is still useful. The total weight needed went down sharply after 1935 when I came on the market with northern goose down.

"Before I invented the quilted down jackets and sleeping bags, I carried up to 120 pounds on a packboard. Everything was heavy. The woolen underwear was heavy, the woolen pants and mackinaw coats were heavy. Sometimes we wore sheepskin shearling-lined coats, and they were heavy too. They would take on water, and it was all you could do just to carry the clothing you were wearing.

"In addition to the heavy clothing I wore, I would take along a waterproof groundcloth and horse-blanket safety pins to pin the blankets together. Now that 120-pound load I carried wouldn't weigh any more than 70 or 75 pounds with the dehydrated foods, down clothing and sleeping bags, and ultralight fabrics.

"1. A lightweight .30-caliber rifle with auxiliary chamber for reduced loads, plus ammunition. It always went with me on treks when there might be a need for food or the chance meeting with potentially dangerous creatures.

"2. A short-jointed 4-oz. flyrod, reel, line, and terminal tackle. In a survival kit, rod and reel are omitted.

"3. A Hudson's Bay ax with leather head sheath, 20- or 24-inch handle.

"4. A belt knife in leather sheath with a 5-inch blade, plus a small sharpening stone to keep the edge keen.

"5. A Silva Polaris compass and topographical map of the travel area.

"6. A small waterproof matchbox for the old-fashioned wood matches. I always carry them on my person for emergencies.

"7. A small folding, or otherwise compact, wood saw. There are several small styles to choose from.

"8. My 1920 and 1940 homemade wooden-frame two-and-one-half-pound packboards. These have served me and friends all through the years carrying loads upwards to 150 and 200 pounds. Today aluminum packs do equally well or better. Many are on the market.

"9. A Norwegian Bergen metal-framed rucksack is ideal for short trips, and convenient with its four side and top-cover compartments. Your sleeping bag rides outside.

"10. My shelter tent which I made in 1915 of unbleached muslin sheeting, seven by twelve feet, and waterproofed with paraffin melted in benzine. It sets up as a half-pyramid for two, as an A-tent for several, or may be used full-sized as a lean-to shelter. The first one was replaced after twenty years with one made of balloon cloth, and I have used that one all through the years.

"11. Bedding: I have managed, under all conditions in temperatures from moderate to minus 40 degrees Fahrenheit, with a variety of bedding, which I will describe by the years used.

a. 1914 to 1935: Temperature fall to 20 degrees without fire for warmth, and down to minus 10 degrees with a campfire reflecting warmth toward my homemade tent: **Two all-wool trapper blankets.** Probably 90 percent of all outdoors people used blankets, although a few did invest in sleeping bags insulated with Kapok wool batting or down.

b. 1935 to 1980: I have used one or both of my trapper blankets periodically but usually a sleeping bag of my own manufacture. I have three of these accumulated through the years for these uses:

Backpacking: Comfort range from moderate to minus 40 degrees. A goose down mummy bag.

Car or packhorse: Moderate to minus 50 degrees. A heavy-duty 90-by-90-inch goose down rectangular bag.

Car or boat: Mild to 20 degrees. A rectangular sleeping bag insulated with three pounds of DuPont Dacron, plus a washable, removable flannel liner. I have used this bag so much during the past twenty years that the nylon outer fabric has worn through from my whiskers.

"12. Miscellaneous: These are periodically useful articles that have been in my inventory as long as sixty-five years, including:

a. An army-type trench shovel, lightweight and about two feet long. The blade is about 6½ inches wide by 7½ or 8 inches long. It is very useful around camp.

b. An ultralight air mattress, 44 x 28 inches. It is excellent for backpacking, and useful both as a bed and flotation device. When weight is less important, one can select larger, more comfortable mattresses.

c. Waterproof poncho (eight feet, seven inches long by 47 inches wide). It weighs 23 ounces and slips over the head to shed rain over pack and person. It has many uses around camp, boating, and horseback.

d. Campfire grid, 20 to 24 inches, with folding legs, plus a lightweight fabric bag or cover.

e. Plastic water bag and water flask. The bag holds up to three gallons, and rolls to pocket size. Flask is pint-sized and I carry it on my person.

f. Vacuum bottle. I carry two pint-sized.

g. Flashlights. I have several to choose from and select the ones that may be useful for specific outings—two-cell pen type, two-cell medium sized, and a large two-celled.

h. Plumber candles. One or more can be useful at times.

i. First-aid kit. No larger than necessary for personal needs. Mine includes safety pins, needles, buttons, and thread.

j. Cooking kit and tableware. Aluminum nesting pots, with or without lids, 10-quart and 2-quart to use on the grid, over hot coals, or to hang over the fire. Take only the sizes needed.

Nesting tin cups, stainless-steel forks and spoons, one or two stainless-steel plates with straight sides, one sheet steel fry pan. Again, take only those needed for the trip.

Useful at times: Paper plates, plastic plates, paper cups, plastic cups and bowls, plastic spoons and forks when going light.

Knife: I use my belt knife.

k. A few nails, assorted sizes, common 6-, 8-, and 10-penny. You'll find them useful.

l. An 8-inch file, flat mill bastard type.

m. A 100-pound pulley. Ultralight, compact, and useful in moving or hanging heavy objects.

n. Braided cord and wire. Both are useful in camp and on treks. I take drapery cord, cotton or nylon, and baling or galvanized telephone wire, as much as I think will be useful.

o. A strong "S" link, closed on one end, to use with wire for cooking to hang a pot or kettle from a tripod of poles over an open fire.

p. Slingshot and rabbit snares, to take small game when necessary. I have always made my own, but now slingshots with an effective range of 225 yards, and weighing seven ounces, are available.

"And finally, always have in mind the possibility of getting wet to the skin by slipping and falling into a stream or in a downpour of rain. It is wise to keep food and matches in waterproof bags or containers. Also keep in mind the windchill factor and carry a windproof parka or substitute garment of ultralight backpacking weight."

SURVIVAL KITS

Most recreation stores stock survival kits that contain important items. Some are in cans no larger than pipe-tobacco cans; others are larger and contain more sophisticated items. A typical kit will have the following items:

- tube tent
- whistle
- candle
- matches
- duct tape
- aluminum foil
- candy bars for energy
- bouillon cubes
- dextrose cubes
- herb tea bags
- signal mirror
- aid supplies
- fire-starter
- nylon cord
- razor blade
- waterproof survival instructions

CAMPER'S SECRET

Consider going with more experienced campers, preferably good friends or relatives, before striking out on your own. Camping isn't as complex as it sounds, but after camping with experienced people, you will have a better idea of what to expect when you go it alone.

Another popular item in many such kits is the "space blanket" which is extremely light and compact, yet reflects up to 80 percent of your body heat when you wrap yourself in it.

As with all basic equipment lists, neither the Ten Essentials nor survival kits are carved in stone, and each camper will have his or her own items to add. Boaters have their own particular needs, as do private pilots. The Coast Guard, Federal Aviation Administration, and search-and-rescue associations should be consulted for their lists.

As an aside, the Tacoma (Washington) Mountain Search and Rescue Council spent months designing a survival kit that would fit inside a tobacco can. The kit has an eight-foot tube tent, a whistle, a signal mirror, matches, a candle, wire, bouillon, tea bags, sugar, salt, and instructions for using the can as a cooking container.

After long discussions, the council members decided to include the tea bags—not for their food value, but simply to give the stranded or lost camper something to do. Decisions made early in an emergency situation can easily be the most important ones, and the council reasoned that if campers had something to do for the first few minutes—namely, to build a fire and wait for the water in the tobacco can to heat enough to brew a cup of tea—they would calm down and be better able to handle the emergency.

Additional emergency equipment for automobiles and RVs is listed below, along with other reminders.

- Check insurance premium dates; pay ahead for period of trip.
- If traveling with pet, make sure you have pet vaccination records.
- Check vaccination dates; get shots if expiration occurs during trip.
- Check credit card expiration dates; renew if needed.
- List credit card numbers on sheet stored away from cards.
- Lock all doors, including garage, and all windows.
- Turn radio on; tune to twenty-four-hour station and turn up volume.
- Make sure you have driver's license, vehicle registration, spare keys.
- Store valuables in a safe place.

A SPECIAL NOTE

Most adults will deny it, but nearly all of us have the equivalent of a security blanket without which we feel badly equipped for a trip. It may be totally useless, but we like to carry it. Children are more open about their attachment to favorite belongings, such as a particularly disreputable stuffed animal with all the fur worn off, a favorite book, or a toothbrush that looks as though it was beaten with a hammer. The point is that whatever makes you feel comfortable, and doesn't take up a lot of space, should be taken along.

So many personal items are taken for granted at home that it is wise to make a list of what is absolutely necessary, such as a pair of reading glasses or a magnifying glass for threading needles or reading minute instructions on food packages. You may also need a paperback book, a sketch pad, or similar items for passing away those hot, quiet afternoons when you decide that relaxation is a major part of your vacation. Remember, going on camping trips does not mean you must rough it completely and shed yourself of everything you enjoy.

ROPES

Ropes (or lines, depending on your choice) are one of the most indispensable parts of your equipment list. You will find more uses for rope than you can imagine while sitting in the family room planning a camping trip. You'll need it for stringing tarps, for repairing tents or reinforcing them, to lash gear onto the family car, as a clothesline, for hanging food from a tree branch, for crossing swift streams, to replace broken boot or shoe laces, or for learning to tie knots on slow afternoons.

For the heavier work, such as hanging the camp larder from a tree branch or even to pull your car from the mud on a small winch, a sixty-foot length of 5/16-inch nylon climbing rope is usually sufficient. This takes little space and weighs less than two pounds, yet has a 2,800-pound capacity.

For lighter work, repairs and so forth, pick up a 100-foot length of the so-called clothesline rope, the small nylon rope you find in virtually every hardware store and some grocery stores. You will find uses for this on every outing.

Remember that when you cut off a length of nylon rope, you should hold both severed ends to an open flame to melt the individual strands into a mass so the rope won't unravel.

Manila and other natural-fiber ropes should be tied off on the ends with heavy thread.

LIGHTING

Even the pioneers of America had artificial lighting for their prairie schooners and cabins, but the state of the art has improved considerably. Now we have everything from the old dependable "gas" lanterns to battery-powered fluorescent lights that can be recharged after each camping trip.

For many campers, the old stand-by is the white gas lantern with the glass globe, the delicate mantles, and mild hissing noise that is as much a part of camp at night as shooting stars.

Relatively new on the market is a much smaller lantern that fits neatly atop a butane cartridge and starts with the flick of a starting switch, something like a cigarette lighter. These are becoming more and more popular because they are lightweight and occupy little space. Also, since the butane cartridges are disposable, you don't have to worry about spilling the fluid when filling it.

Also relatively new is the battery-operated fluorescent light, perhaps the brightest of the camping lanterns. It can either be operated from the rechargeable battery or plugged into a 110-volt outlet (which also is used to recharge the battery).

Flashlights: A bewildering array of flashlights are on the market, from the tiny models that hang from your keychain to the 12-volt models that are almost strong enough to double as a search light. Some are disposable; when the battery goes out, the whole thing goes in the garbage.

CAMPER'S SECRET

When shopping for a camping flashlight, consider buying one that has a clip inside for a spare bulb. These cost very little more, and you'll never have to dig through the packs for a spare bulb that was probably left at home anyway.

LANTERNS

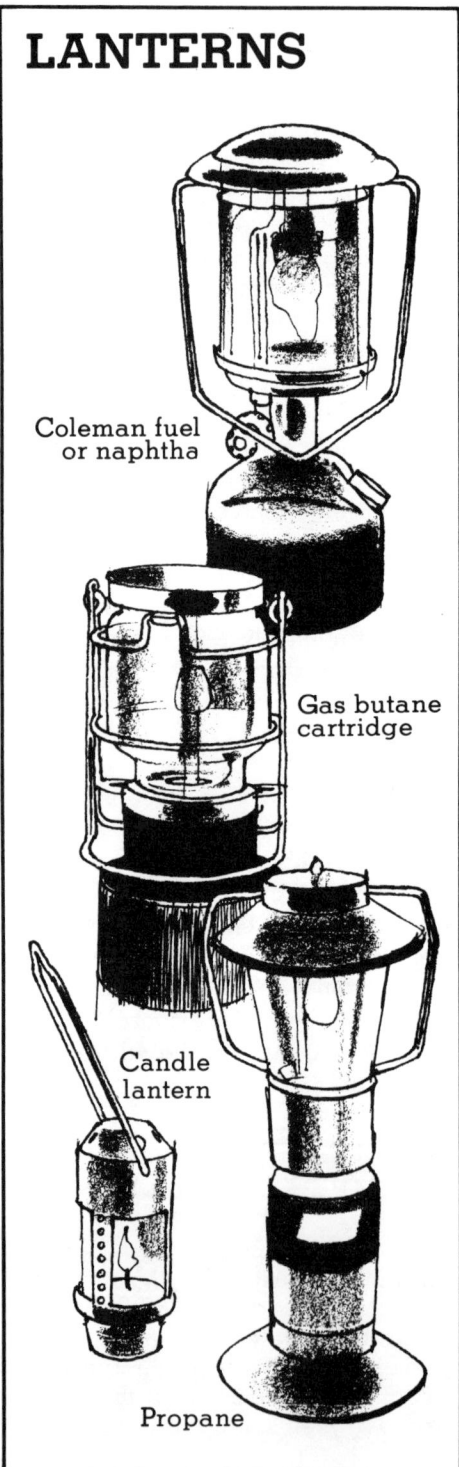

Coleman fuel or naphtha

Gas butane cartridge

Candle lantern

Propane

The best for most purposes are still those that have the main elements of lens, bulb, and batteries, all of which can be replaced. The important thing, of course, is reliability. When you wake in the night and go stumbling off toward the latrine, you do not want a flashlight that abandons you halfway there.

The best models are waterproof, and many of the plastic ones will float should you drop them into a lake or stream.

The most practical models have a clip inside to hold a spare bulb, the one part many people frequently forget to carry.

Kids (and some adults) are addicted to playing with flashlights, even at high noon, so that by the second night of camping, you often have at the best a weak yellow light. Whether this is the case with your group or not, always keep a spare set of batteries tucked away out of sight. And hide the flashlight at dawn if you have to.

A number of small candle lanterns are available, which are simply metal cases for candles with transparent lenses. These are obviously much safer to use than plain candles and have the advantage of being windproof.

Another convenient light is the adaptation of miners' lanterns that you slip over your head. There are battery-powered and are much more convenient than hand-held flashlights because they free your hands and point in the directions your head is turned. These also are ideal for the camper who can't face a night in bed without reading a chapter or two of a novel.

Candles: Also available are small candle lanterns, a small cylinder in which a candle is placed. These can provide enough light in a tent so you can tell the difference between a sleeping bag and a parka, but few people can depend on

CAMPING EQUIPMENT—THE ESSENTIALS

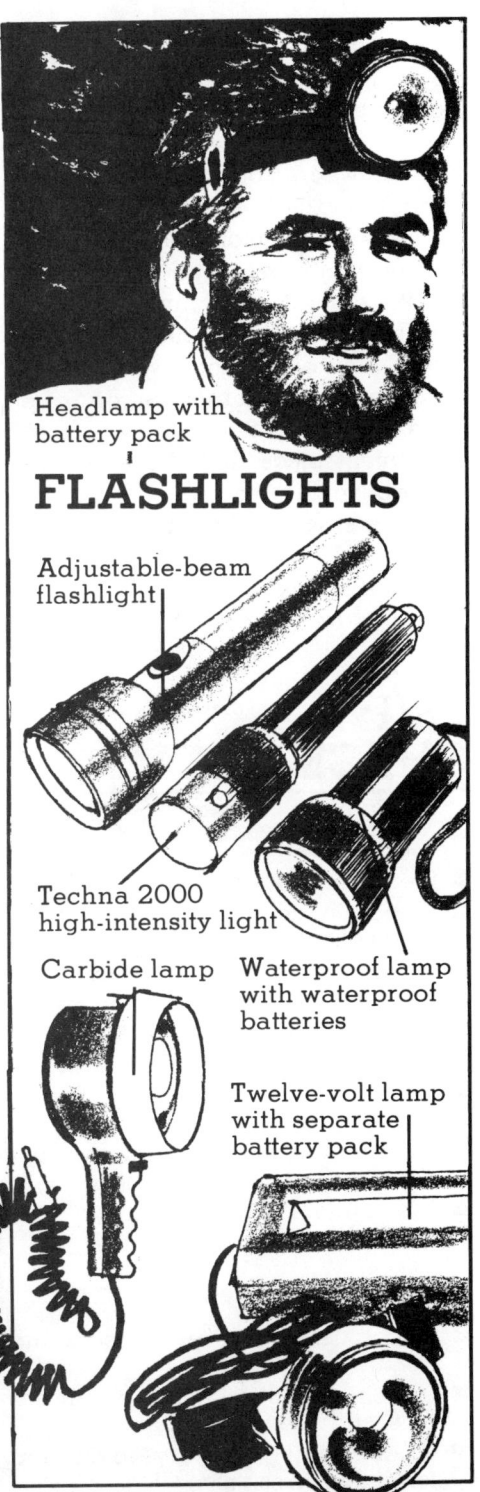

FLASHLIGHTS

Headlamp with battery pack

Adjustable-beam flashlight

Techna 2000 high-intensity light

Carbide lamp

Waterproof lamp with waterproof batteries

Twelve-volt lamp with separate battery pack

them for nighttime reading. Their main value seems to be psychological—a small ray of light in the blackness.

PACKING

Few things are more frustrating to the beginning camper than trying to cram everything you think you need on a trip into the family car that you normally think is crowded when the whole family goes to visit a relative on Sunday afternoon. The key phrase here is **everything you think you need,** because no matter how long you've camped, you will want to bring more than is actually needed.

This is one of the best arguments for buying the newer ultralight and compact camping equipment and freeze-dried and dehydrated food now available. If you own a medium-sized car—or even a sub-compact with average trunk space—and you avoid packing breakables, you can put a medium-sized ice chest, camping stove, cooking utensils, food, two tents, clothing, six sleeping bags, and a few other items in the trunk. Other smaller items can be stuffed in nooks and crannies beneath the seats, at your feet, and along the rear window.

If at all possible, avoid using cartop carriers for anything other than bicycles, skis, and so forth, because the enclosed cartop boxes are brutal on gasoline mileage. However, if you're forced to cram six people, all their equipment for a week, plus a pet or two into a four-person vehicle, then you have no choice but to use a car top carrier.

The best means of carrying your own gear is in a backpack with all its compartments for ease of separating items by their use: sleeping bag pad and tent attached to the bottom, toilet articles in one side pocket, eating utensils in another, camping tools in still another, and so forth.

The difficulty of using several backpacks with frame intact is that they are often difficult to arrange in the car trunk for a party of more than three. If you know you won't be going on anything more strenuous than a day hike, you can remove the pack from its frame.

If you don't own backpacks and don't plan to use them, the next best alternative is a duffel bag for each person in the group. The longer the zipper in a duffel bag, the better. One of camping's little frustrations is rummaging around in a duffel bag in search of something on the very bottom when the zipper is at one end. The best choices, like packs, have more than one compartment so you can divide the gear into easy-to-find places.

Obviously you won't want to take anything along that will break easily, no glass objects, no delicate wooden models, and as few items with sharp edges as possible.

When everything is stacked beside the car ready for stowing, run an inventory to be sure that en route snacks, rain gear, reading material, and car games are not in the group that will be stowed in the trunk. And have a plastic bag or grocery bag in the car with you for the amazing amount of debris that accumulates between your garage and the campground.

Stow the unyielding objects, such as coolers, campstoves, etc., first and then stuff the softer items around them, such as sleeping bags and clothing. Keep the cooler accessible since you may be stopping for a snack or soft drinks en route. Also keep the car's tool kit in a convenient place. It is better to stow things beneath the passengers' feet than on the rear window shelf because it is both illegal and dangerous to obstruct the driver's view.

TRIP CHECK LIST

Vehicle Preparation
- Get engine tune-up.
- Check battery level and battery connections.
- Check tire pressure and tread; check spare.
- Check transmission fluid.
- Check radiator level.
- Check all lights—interior, brake, backup lights, etc.
- Check trailer hitch for tightness, connection points, and hitch ball.
- Clean inside of windows.
- Check tool kit for all essentials.

RV Preparation
- Check trailer tires for pressure and tread.
- Check all trailer lights and turn signals, and interior lights.
- Check trailer battery level and terminal corrosion.

CAMPER'S SECRET

Use every available inch of everything for packing. If you're carrying a two-burner stove, you can pack most of your cooking implements and perhaps your dishwashing soap or other items in it. Do not put food in a campstove, since it may well have a residue of fuel or soot.

CAMPING EQUIPMENT—THE ESSENTIALS

- Check all air vents for blockage (furnace, refrigerator).
- Fill water tanks; test water pump.
- Check gas-light mantles.
- Add chemical to holding tanks.
- Make sure power cord, water hose, jacks, and disposal hose are aboard.
- Close windows and roof vents tight.
- Check and top off propane tanks.

Travel Information
- Prepare detailed route plan.
- Identify potential scenic areas, recreation stops, other "don't miss" places.
- Estimate travel time and time of arrival.
- Make and/or confirm reservations.
- Advise campground of any delays.

Recreation Gear
- fishing tackle, rods, and reels
- camera, film, accessory items
- binoculars
- tennis rackets
- balls (volleyball, softball, football)
- Frisbee
- swimsuits and swimming gear
- games, cards, toys, car toys
- life jackets

Kitchen Gear
- can and bottle opener
- heating kettles or assortment of pans and kettles
- stirring and serving spoons and forks
- eating utensils
- pancake turner
- wiener or marshmallow cooking forks
- hot pads
- plastic tablecloth
- serving platter
- coffee pot
- pitcher
- sealable plastic containers

Camping Gear
- tent, poles, stakes
- ground cloth
- sleeping bags
- cots, air mattresses
- lantern
- stove
- catalytic heater
- cooler
- charcoal grill
- folding chairs
- folding table
- ax, saw
- water jug
- flashlight
- stove stand
- lantern stand

Securing Your Home
- Notify police of exact dates of your trip.
- Stop mail delivery at post office or have mail picked up by a neighbor.
- Stop deliveries (paper, milk, etc.).
- Arrange for someone to mow grass or shovel walks while you are away.
- Arrange for a friend or neighbor to turn on different lights every few days.
- Arrange to have plants watered.
- Empty refrigerator of perishable foods.
- Unplug all unnecessary electrical appliances.
- Turn off gas to kitchen range.

CHAPTER 4

TENTS AND SHELTERS

Tents are your home away from home, and you should exercise the same care in buying one you would when buying a home or vacation home. Under normal circumstances you won't spend many waking hours in a tent, but should your camping trip coincide with the local rainy season, you will become intimately familiar with the interior of the tent. So it is important that you choose one in which you will be comfortable.

The first generation of tents tended to be rather dim and confining—like a house with all the shutters drawn. In spite of their having windows, most tended to be dark and dismal even on a bright day. But the newer fabrics and better designs have turned today's tents into light and airy dwellings. The walls of solid canvas have been replaced in some designs with a fine mesh that lets both light and air inside—and yet keeps bugs out. At night the flaps can be unrolled to cover the mesh, giving privacy and comfort.

Gone, too, are the days when all tents were so heavy it took at least two people to carry one, when leaks were taken for granted and mildew a fact of life. Today there is an almost bewildering variety of tents of all fabrics, designs, and weights. Although the large wall tent is still popular with campers who stay within sight of their vehicles or who use pack animals, there are roomy tents available so light that a six-year-old can carry one.

Most tents on the market today are made of lightweight, sheer nylon. The basic tent is intentionally not waterproof on the top and sides, whereas the floor and perhaps a few inches up the walls are coated with waterproof material to keep groundwater out. The walls and top are made of breathable nylon material so that moisture will pass through rather than condense on the walls and ceiling. Strung over the entire tent is a rain fly, usually nylon coated with waterproofing. The fly is supported a few inches away from the tent to allow air flow that carries away the moisture that comes from bodies and breathing inside. The rain fly also protects the tent against damage from ultraviolet rays.

Nearly all tents are equipped with a fine-mesh screen over all doors and windows to keep out insects. A common description of this mesh is "noseeumproof" because the irritating little bugs called no-seeums, plus gnats, chiggers, mosquitoes, flies, and bees, are the bane of older designs without this durable netting.

In addition to this netting, which is usually zippered at the entrances and either permanent or closed with drawstrings at the windows and vents, tents also have flaps that cover the entrances and windows for privacy and comfort.

Because they are porous, very few tents, including the canvas models, are designed to keep you warm. But they do protect you from rain, heavy winds, snow, and sun. In extremely cold situations, such as climbing expeditions or winter steelheading and cross-country skiing, you can use tent heaters to keep much of the cold out, but carefully observe all safety precautions if you use them. For the most part, you must depend on your clothing and sleeping bags for warmth.

Some of the classic canvas wall tents have zippered and fireproofed openings in the roof or high on the walls for stovepipes so you can use a wood-stove and have the coziness of a mountain cabin.

Since it is impossible to find a tent that is good for all kinds of camping, before purchasing one you should decide how you plan to use it. A good rule of thumb regarding the heavier wall tents is that if you never plan to camp more than a hundred yards from your car or boat, or if you will use it on horsepacking trips, you will have no problems. You will have enough space to set up cots, if you prefer them, and you can set up a stove as well.

However, most families expect to do some backpacking, and this automatically rules out the tents weighing more than 12 pounds. You are a candidate for a lightweight tent weighing not more than ten pounds.

Rather than follow the manufacturer's recommendations on how many persons a tent will sleep, you might want to compute your tent needs on a square-foot basis. You can assume that each person is six feet tall and needs a three-foot-wide space for sleeping. Add some room for personal equipment, such as packs and clothing, and you get 27 square feet per person in each tent. For extended periods of camping, such as a week or longer, increase the square footage requirement to approximately 30 feet.

CAMPER'S SECRET

You can rent tents, sleeping bags, and some other equipment before deciding if you want to invest in camping gear. Always test it before leaving the rental store, and be sure you know how to use the equipment.

TENTS AND SHELTERS 59

These figures are only estimates, of course, and you may decide you need less, especially if you have young children or if you are a firm believer in traveling light and plan to crowd three adults in a single tent.

With all these factors in mind, it is well to remember that in spite of all the things you should consider, most campers purchase a backpacking tent weighing less than ten pounds and make it work for all kinds of camping. A backpacking tent weighing six or seven pounds can easily be used for car-camping trips, river trips, extended backpacking expeditions, bike touring, and so forth.

Another possibility is a large tent for adults and a small "pup" type for children.

When shopping for a tent, it is wise to talk to people who already own tents, and, if possible, to rent or borrow a model similar to one you want to buy. In some cases, your individual needs will dictate the design and shape you choose. For example, you may want to connect tents so that children won't be totally separated from their parents, yet can have their privacy, too. In this case, a properly placed tarp or rainfly can be used. This

CAMPER'S SECRET

When you buy your tent, always try to buy a swatch of matching material for repairs and carry it with you at all times. Watch places of extreme wear, such as sleeves for the poles and around the entrance, and when signs of wear appear, make the repairs before the material wears through or tears.

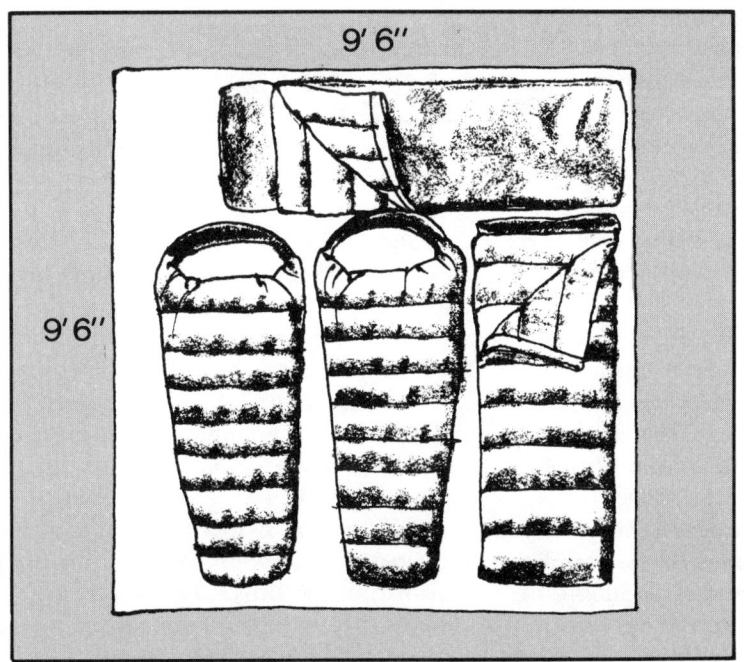

Rather than follow the manufacturers recommendations on how many persons a tent will sleep, you might want to compute your tent needs on a square-foot basis.

system is useful in poor weather, since you can visit back and forth without getting wet or cold.

Many backpacking models come with two carrying bags so that the load can be distributed nearly evenly between those using it: the tent in one bag, and the poles, stakes, and rain fly in the other.

Many families who camp in RVs take along a small tent that will sleep two or three children and pitch it outside the RV so the children can sample sleeping out. Also, children like to get away from adults (and vice versa) occasionally. Such tents are convenient for the hunters and fishers of the family who get up before dawn, then relax with a nap at midday before going out again in the evening.

You will find tents available in a wide variety of colors, and it doesn't seem to make much difference whether you select a vivid or a subdued color scheme. Some veteran campers say it helps to select a bright color in the event you're forced to spend a lot of time inside to wait out bad weather.

The colors range from International orange to purple to blue, green, and white. Some tents have as many as three separate colors, almost like a circus tent. Frequently the floor will be one color, usually a darker tone, and the walls and ceiling a brighter color with the rain fly still another color.

Since tents come in an astonishing variety of sizes as well as shapes, many families purchase tents for use as weekend or vacation homes. These are large enough for a small family and are usually erected on a special wooden platform built for that purpose. The classic American Indian tepee is a popular design for this use and is available in many specialized outdoor stores. Patterns for do-it-yourselfers are also available.

Another type of tent that has been popular in Europe, where recreational opportunities are more limited than in North America, are the very elaborate dome-style tents with private sleeping quarters in the wings and with communal spaces in the middle. As with the tepees, these are erected only when in use and stored at home the rest of the year.

There is enough demand for the standard wall tent that measures at least nine by twelve feet to keep several manufactures in business. These are getting lighter and lighter in weight as cotton duck is replaced by synthetic fabrics. They offer more headroom than the smaller designs that can be used for backpacking and are popular with families who tow a small trailer with camping gear behind their family car.

BASIC TENT MATERIALS

Although research into tent materials and construction continues at a steady pace, during the past few years most manufacturers have settled on nylon as the major material. A very few use Dacron, and some have experimented with various other materials that are lightweight and durable. Cotton, cotton duck, and canvas are used only in the larger tents that are not meant to be carried far and are used for base-camp tents that will sleep the whole family.

Ripstop nylon and nylon taffeta are the most common fabrics found in tents, ranging from the small two-person mountain tents to the larger three- and four-person models that usually weigh less than 12 pounds, rain fly included.

They are of almost equal weight and durability.

Taffeta is a tightly woven nylon that is soft to the touch and permits air and moisture to pass through rather than collecting into condensation.

Ripstop nylon is also tightly woven but has heavier threads approximately three-sixteenths inch apart to prevent tears from spreading. Nylon will eventually break down from ultraviolet light unless the tent's rain fly is kept over it for protection.

Dacron is used by a few manufacturers but it is more expensive than nylon and more difficult to work with. Its major advantage is that it is virtually impervious to ultraviolet light damage from the sun.

BASIC TENT DESIGNS

In spite of the bewildering variety of designs available today, the basic tent shapes are still as follows:

- A-frame, or pup tent
- pyramid
- dome
- wall tent
- "bakers" (lean-to)

A-frame: The first major design was the A-frame or pup tent favored by soldiers and Boy Scouts. It could be erected with a line strung between two trees and the tent hanging over it, or with a single pole at each end. Then each corner was stretched taut and held down by a stake in the ground. Later versions had aluminum or fiberglass poles forming an A-frame at each end so that no poles were inside the tent.

Pup tents are still very much in demand, and the design is constantly undergoing modifications. One disadvantage of the earlier designs was the need of several lines to hold the tent in place, which invariably became tripwires for anyone walking nearby. People who owned the earlier models soon learned to expect frequent shouts of rage from their children as the family clown pulled up the stakes holding down the lines, collapsing the tent onto the inhabitants.

Today many tents of the classic A-frame design have no exterior lines. Most avoid this by having stakes all around the edge of the floor and a tension bar across the top holding the A-frames at each end taut and erect. Various devices have been created to keep the rain fly a few inches away from the tent wall so that the tent will "breathe" and avoid condensation on the walls.

Pyramid: Like the A-frame, this is a basic—and classic—design that has been used for centuries by Mongols, Lapps, and North American Indians. So basic is the design that its modifications during this century have been relatively minor.

Early canvas tents of this design, however, often depended on a single center pole that had a disquieting tendency to collapse on the occupants, or always to be in the way. The next step, after aluminum and fiberglass poles became common, was to insert three or four lightweight poles in sleeves sewn into the cover, thus removing poles from inside the tent.

This taller design is particularly popular with campers who find it difficult to pull on their trousers without standing up. Many models have tabs sewn into the walls about three feet up so that the walls can be stretched out, giving occupants a bit of additional sleep-

Umbrella

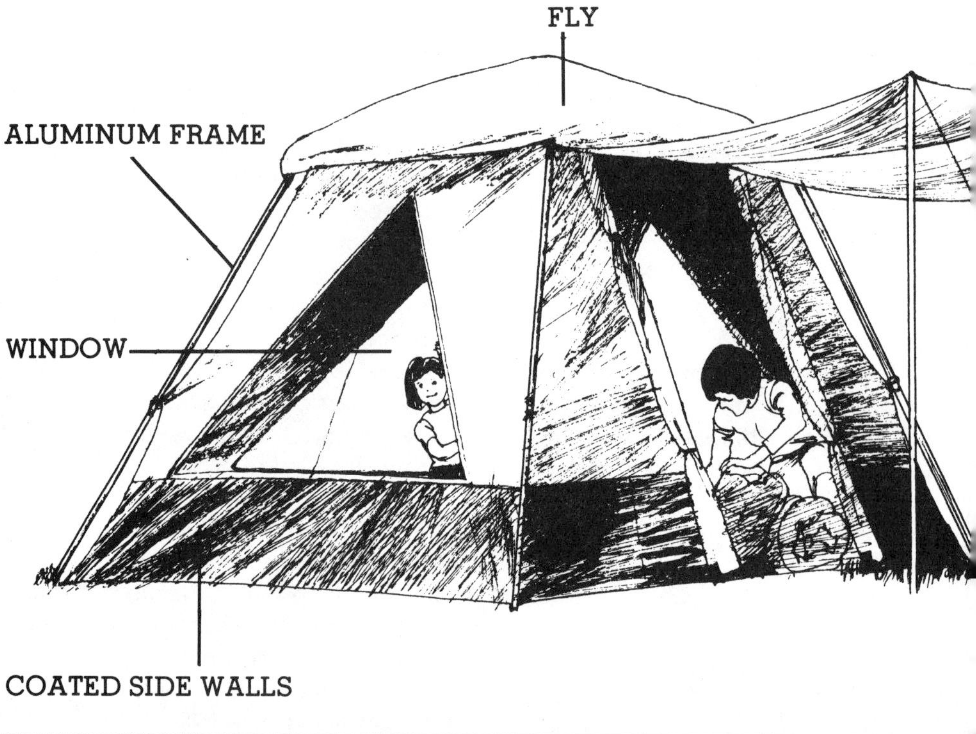

- FLY
- ALUMINUM FRAME
- WINDOW
- COATED SIDE WALLS

A-frame

TENTS AND SHELTERS 63

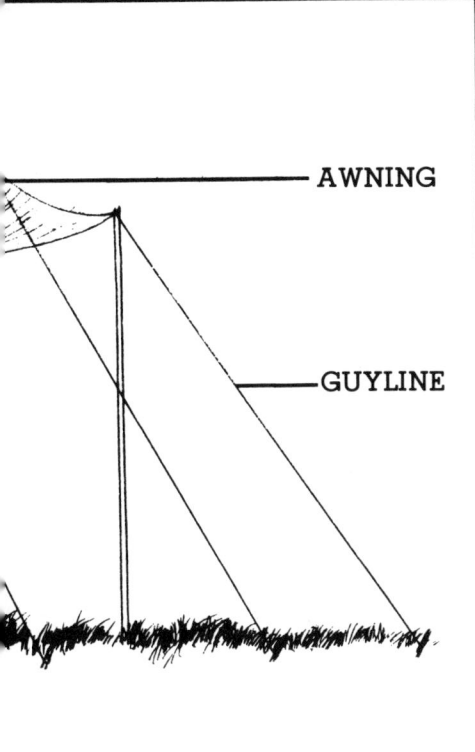

TENTS

Some things to watch for in tent construction include:

1. Reinforced pole pockets, preferably webbing, so the poles won't rip away from the tent.
2. Catenary cut on ridgelines so tent can be kept taut.
3. Noseeum-proof netting over entrances and vents.
4. Nylon coil zippers.
5. "Tub" floors that are waterproof part way up walls.

CAMPER'S SECRET

Never go camping with tents you haven't put up at least once at home. Always pretest them so you can cut down on the time it takes to get them erected, and to inventory them to see which pieces are missing.

Dome

ing space without having the walls against their sleeping bags or faces.

Some pyramid-style tents are variations of the design, such as a half-pyramid with the front open to catch heat from campfires reflected inside. This design permits you to erect two half-pyramids facing each other with a rain fly or tarp stretched over the openings to give you a weather-free opening. An example of this versatile tent is Eddie Bauer's first tent, shown on page 43, one he designed for himself and later sold through his stores.

Dome: This design is the newest on the market, and there are perhaps hundreds of variations from which to choose. Although many people buy a dome simply because it is more attractive than other designs and reminiscent of Buckminster Fuller's designs or Canadian Eskimo igloos, this tent has almost an equal number of advantages and disadvantages.

First the advantages: It is self-supporting, meaning that no stakes or lines are required under normal situations. This makes it simple to clean; just empty it, pick it up, and shake out the dirt and sand. It is more aerodynamic than most tents because it offers no flat surface for the wind to strike and whip the tent.

One disadvantage is that the dome can be hard to set up, especially in the dark or with the wind blowing, when it threatens to become a kite while you are inserting the poles through the sleeves that encircle them. The dome also tends to weigh more because more poles are required.

Although no stakes or lines are required to keep dome tents erect, you should never leave them unoccupied without staking them down. They are so light that even a slight breeze can send an empty dome tent tumbleweeding off into the wilderness.

The most satisfactory models seem to be the modified dome tents, those which use only two poles and have flat sides. Because they have fewer poles, modified dome tents weigh substantially less and they are easier to erect in all situations.

However, you should remember that no single tent is all things to all campers. The disadvantages of the pure dome tents do not bother most campers who own them. They become quite adept at erecting them even in a midnight gale and would not consider owning another style.

Wall tent: This is a classic design that was used well before the turn of the century by miners and hunters and is still preferred by campers who set up a base camp and never move from it, such as hunting parties or groups spending an entire vacation at one site.

The walls of these tents come up four or five feet before forming a roof. Most are made of cotton or canvas, although some of the new lightweight materials are finding their way into this design. A system of poles and stakes keeps the walls and top taut. Some models require guy lines, but most do not. Some also have awnings attached to form a porch.

A few camping outfitters build half cabins with floors and use the wall tents as a covering and roof, taking the tents down at the end of the season. Wall tents are actually a temporary cabin, and you can set up housekeeping for a week or a month, complete with ice chest, cookstove, bunks, and lanterns.

Some available models of wall tents are designed especially for owners of vans. These have a vestibule that connects the tent with the van's door and are

TENTS AND SHELTERS

held to the van with padded magnets or other devices.

Baker tents: These wall tents have a roof that slopes backward from the front with a flap that can be used as an awning or porch during the day and dropped as a covering flap at night. They are one of the simplest to erect, though not recommended for camping where bugs are a problem, unless you equip the tent with netting or add the equivalent of a screened porch with insect netting.

BUYING A TENT

It isn't a law of the outdoors, but in most cases the less an item weighs, the more it costs. Although weight isn't so important for regular family or RV camping near your source of transportation, you will be wise to keep it in mind when buying tents and sleeping bags. Will these items ever be used for backpacking or bicycle touring? Will they be used by adults only, or will teenagers, who may be more careless than adults, be using them for group outings, slumber parties, and the like?

If you are planning to make these items last for years and years, you should consider buying the top-of-the-line models.

Choosing the material, design, color scheme, and other factors of a tent is at first an overwhelming task to the first-time buyer. But a few outings in different styles of tents will help you narrow the choice, so the best way to determine your needs is either to rent a variety of tents or to go camping with friends and listen to the pros and cons of each style.

In addition, be sure to consider the following before purchasing your tent:

> **CAMPER'S SECRET**
>
> If your tent needs guy lines to support it, one way to help avoid the problem of people tripping over the lines day and night is to hang pieces of white plastic or a similar material to the lines. These "flags" will show well on moonlit nights as well.

Size: How many people will sleep in it? Will it be too large for some campsites? How often will you move it on each outing?

Weight: Will you use it only for car camping, or will you be using it for backpacking, canoe trips, bike touring, and other forms of travel when weight is critical?

Design: Do you require lots of head room? Do you want a self-supporting design or one with guy lines? Will you be camping in the winter and require a vestibule, cook hole, and tunnel entrance?

TIPS ON TENT CARE

Storage: Before storing your tent after use, be certain it is totally dry and clean. Erect it in your backyard, carport, or family room and sweep out all dirt, sand, and other foreign matter you may have left in when breaking camp. Turn stuff or storage bag inside out to clean and dry. Wipe all poles and stakes. Do not leave the tent in sunlight longer than necessary or ultraviolet rays will cause the fabric to deteriorate. When it is out to dry, erect the rain fly also since it is less susceptible to ultraviolet rays. Do not

66 FAMILY CAMPING

Always have the salesclerk show you how to erect your tent, then practice erecting it in your backyard or family room.

TENTS AND SHELTERS   67

leave the tent in a car trunk or other hot place since high temperatures can damage the fabric, too. Store in a cool, dry place in its storage bag.

Leaks: Buy seam sealer, available at outfitters, and use as directed, even on new tents, because the majority of leaks occur in the seams. Apply sealer as needed throughout the tent's life.

Cleaning: Use either a very mild liquid detergent or a baking soda and water solution. Be certain no residue is left on the fabric. In all cases, follow the manufacturer's or dealer's recommendations.

Staking: More and more tents are self-supporting and have no lines to stake down. However, always take aluminum or hard plastic stakes and use them as anchors to avoid watching the tent roll away like a tumbleweed in a strong wind, sleeping bag and all.

Poles: Keep the poles and fiberglass wands wiped clean and free of corrosion. If a burr develops on a metal pole or metal connector for fiberglass wands, buff it down with fine sandpaper or a nail file. Otherwise, a hole in the fabric will develop very quickly.

Shock cords: If your tent is not equipped with shock cords around the base and the rain fly, it is wise to install them because the tent fabric will expand and contract with heat and humidity changes. These cords, made of heavy-duty elastic, are available in various degrees of stiffness and can be cut to the desired length.

Zippers: Keep the zippers coated with a light silicone lubricant to prevent freezing or snagging. Stick or liquid lubricant can be purchased at most camping equipment stores.

Familiarity: Always understand clearly how to erect the tent you buy or rent, and have the clerk show you how before leaving the store. Practice putting it up at least once before going on a trip to be sure all the parts fit. A prime source of frustration is trying to erect an unfamiliar tent in darkness, wind, or rain.

Chemical problems: Be wary of using chemicals such as hair spray inside your nylon tent. Hair spray can damage the fibers.

Fire safety: Most tents manufactured since the mid-1970s are treated with chemicals to render them fire-resistant. This does not mean they won't burn if exposed to an open flame. But it does mean that they won't turn into an instant torch as some earlier models did. Always check the specifications of a tent before purchase to be certain it is treated with a fire retardant. It is required by law in most states, and nearly all manufacturers comply with these laws in all their tents.

Obviously you should not cook in your tent unless absolutely necessary, such as when you're winter camping and caught in a blizzard. Some tents designed for winter use have a zipout cook hole in the floor so that spilled fuel and food will go into the snow and the tipped-over stove will fall harmlessly.

CAMPER'S SECRET

If your tarp doesn't have metal grommets, you can make temporary ones by wadding a small handful of moist dirt and wrapping a corner of the tarp around it. Tie the lump with a line, and you have a grommet. Of course, it is best always to have a few ball-and-wire grommets with you.

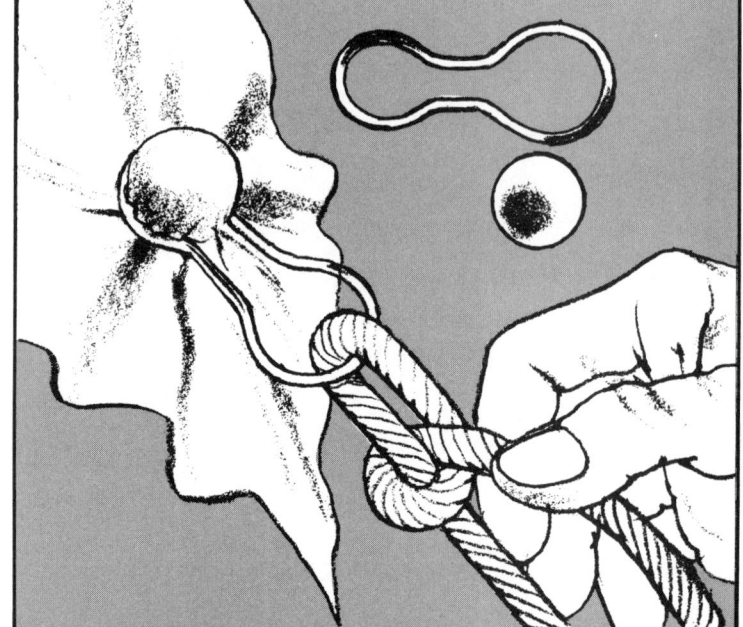

Ball-and-wire grommets are simple to use. You should carry one or two extra with you in case you want to use a poncho for a temporary shelter or in case a grommet in your tent tears.

OTHER SHELTERS

Many experienced campers, particularly those who go on long backpacking trips in warm weather or on river trips, carry only the minimum in shelter: a simple waterproof nylon tarp or plastic sheet about nine by twelve feet. They use this for making either a simple lean-to at night or an open-ended pup tent.

Another choice is the tube tent, which is little more than a plastic sack a person can crawl into, usually open at both ends so it can be suspended above you with a piece of rope stretched between two trees or bushes. Some campers buy an inexpensive plastic tube tent to stake out a campsite while they drive their car or RV elsewhere during the day.

For most car campers, these shelters are barely adequate and are carried for emergency use or as an adjunct to regular tents.

CAMPER'S SECRET

Nearly all tents and lines have a tendency alternately to shrink or to sag. One way to help adjust this condition is to attach a loop of shock cord at the end of the lines where they connect to the poles or stakes. Shock cord is a sturdy elastic that can be purchased in small loops or by the foot. Many tent poles that are hollow and must be fitted together can be kept together by running shock cord through them with tension on them when the poles are together. When it's time to fit the pole sections together, give a snap of the wrist and they pop into place.

CHAPTER 5

BEDDING

Until Eddie Bauer invented quilted goose down-filled garments and sleeping bags in the mid-1930s, campers had to rely on sleeping gear that was only a little lighter and less bulky then tents. They carried blankets or bags stuffed with various battings that were much like homemade quilts.

The modern sleeping bag made of ultralight nylon filled with goose down didn't come into its own until World War II, when Eddie Bauer manufactured them for Army Air Corps crewmen and mountain troops.

The first generation of sleeping bags without goose down simply did not keep campers warm. They often had to take turns getting up at night to stoke the fire in freezing weather. Sleeping in hunting cabins was a bit more comfortable, but in subzero weather the woodstove often had to be kept glowing through the night.

With today's bags, you can sleep in comfort even on the world's tallest mountains without having enormous layers of wool weighing you against the ground. The difference is the down bags and insulating pads beneath you to prevent the conduction of ground cold upward.

Campers also have another new generation of synthetic insulation from which to choose. This insulation, which comes in a variety of kinds and is sold under brand names such as Polarguard, Hollofil, and Thinsulate, is gaining popularity for use in both sleeping bags and garments. Although it is not an across-the-board substitute for down, it is being improved constantly and has a significant niche in the outdoor market.

These synthetics are made of polyester and are manufactured in sheets or batts, then cut to size for each garment or sleeping bag. These batts create millions of tiny air pockets to provide loft, as in goose down, where the dead air creates an insulation barrier.

The advantage of these products over goose down, besides their lower cost, is a resistance to absorbing moisture so that they retain their loft even when wet. If a synthetic-filled item becomes soaked, it can be simply wrung out and the fill springs back to its original shape. Conversely, once saturated, goose down does not regain its loft until it is totally dry, and wringing it out compresses the down further.

The major disadvantages of synthetic insulation are that it requires more weight and bulk to provide warmth, and its life span is much shorter than down's. Some synthetic insulations will not last more than three or four years of hard use. Goose down will last several generations with the proper care.

The rule of price and weight is a major factor here: the least weight and the best insulation cost more.

SOME TERMS YOU SHOULD KNOW

Down: Down is the underplumage of waterfowl closest to the underbody. It does not have a shaft like a feather. It has a quill point that a network of arms cluster around, radiating out like an octopus's arms. The best down comes from birds that live in Arctic conditions, where the harsh cold requires millions of the tiny clusters of down for the bird to survive.

The best down of them all, and the most difficult to obtain, is the eiderdown from the Arctic eider duck. This down must be picked from eider nests by hand, and the supply is obviously very limited. The difference between this and regular goose down is that eiderdown has tiny Velcro-like hooks on it which keeps the down clustered together. Throw a handful of eiderdown into the air and it will cling together in a ball; goose down separates.

However, goose down is the most superior insulation available. Duck down is the second best, and the synthetics come next. A recent study by the U.S. Army's research laboratory in Natick, Massachusetts, states: "In terms of sheer warmth per pound, down is still the best. It's also unmatched in its ability to be compacted without losing resiliency, so you have to say we have no synthetic fibers that will outperform Mother Nature in every respect."

Thinsulate: This is the trade name for one of the newest synthetics on the market. It consists of uniform fibers that have a small crimp in each to hold themselves apart to create the dead-air spaces needed for insulation. Its manufacturer says it requires less loft than down, but also requires more weight.

Loft: Loft is the thickness of a bag or article of clothing when it is fluffed out and dry. The thicker the loft, the better the insulation value. Thus, when you remove

CAMPER'S SECRET

Try to buy different colored or shaped duffel bags and sleeping bags for each member of the family. It will make identification much easier.

BEDDING 73

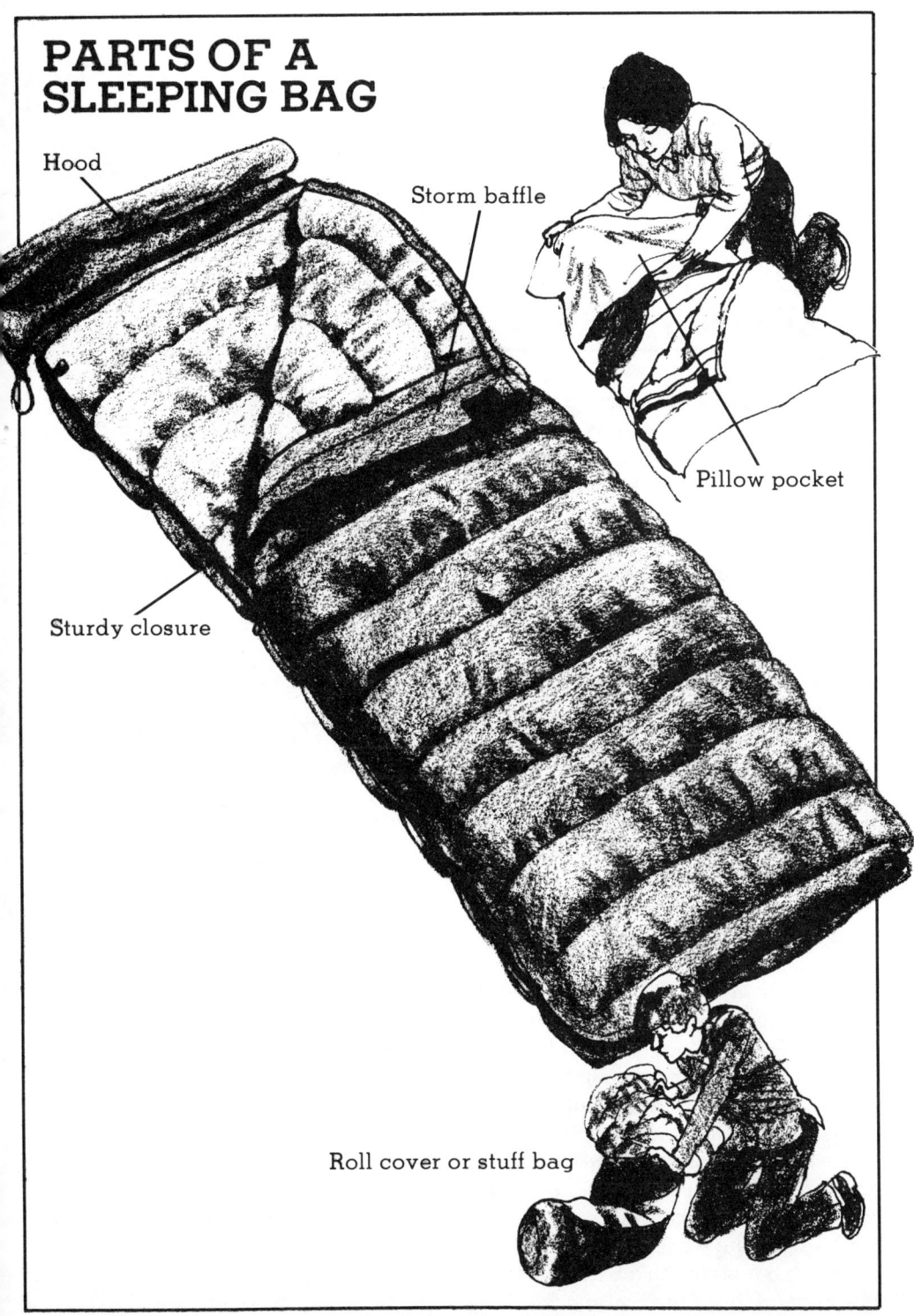

PARTS OF A SLEEPING BAG

Hood
Storm baffle
Pillow pocket
Sturdy closure
Roll cover or stuff bag

a sleeping bag or garment from its stuff bag, you must shake and fluff it to restore the loft, to make all the millions of individual clusters of goose down or synthetic fibers spring back to their normal shape.

CHOOSING A SLEEPING BAG

When shopping for a sleeping bag, there are a number of factors to consider.

First, what will be the nature of your activities? For backpacking, hiking, and bike touring, in the warmer months, you should consider a lightweight bag that you can carry on your pack and that is versatile enough to stretch your season from spring to fall.

For fishing, hunting, boating, and camping, look for a bag that is durable, full-sized, and more comfortable than a backpacking bag since you most likely won't be carrying it for miles and miles.

For canoeing and kayaking, consider a synthetic bag that is both lightweight and compact.

For mountaineering, expeditions, ski touring, and snowshoeing, you will want a bag that is still lightweight, but also the most thermally efficient and durable on the market.

Closely related to the information above is consideration of the temperature range you will be exposed to.

The next consideration is the shape of bag best for your uses. There are two basic shapes— rectangular and mummy, with modifications of each. Generally speaking, the shape you purchase depends on the use to which you will put the bag and the weather conditons. Still speaking generally, on a basis of weight, the mummy bags are warmer because

HOW TO SALVAGE DOWN

Millions of down-insulated items are discarded through charitable organizations such as Goodwill and St. Vincent de Paul thrift shops. Look for the brand names prior to 1968 such as Bauer, Temco, or Comfy. Other products with excellent down are World War II military items such as flying suits and sleeping bags.

Good down will outwear several generations of fabrics, and clothing pattern are readily available in fabric shops. Sometimes you can use the discarded clothing or bags as a pattern if it is your size. Or you can design your own pattern to serve your needs.

Once you find a discarded item insulated with good down, place it in a laundry basin in very warm water with a mild, nonchlorine detergent to both wet and cleanse the down. Launder it the same as other clothing.

Then drain and rinse it, and hand-squeeze it to damp only. This will allow you to remove the down by hand so it can be placed in a suitable "downproof" muslin or ticking bag.

Use these bags for down storage only while you are making your new clothing or bedding. When an item is ready for inserting the down, the wetting procedure is reversed. Wet the down, squeeze out the surplus water, then open the storage bag and remove the down as needed.

By wetting the down, you make it easily manageable and keep it from flying around the house. Water doesn't hurt down; it comes from waterfowl.

It is best to "spank" the down to one end of the container before wetting it. This will keep it gathered into one mass rather than clinging to the entire storage bag.

Fabrics for down products must be nearly "zero" in porosity, or you will be troubled with down seeping through and clinging to other fabrics. The outer fabrics should be water-repellent and durable, such as nylon or blends of long-staple cotton and nylon or polyester, or very tightly woven 100 percent cotton.

The inner fabrics that retain the down need not be water-repellent but should be long-lasting. I prefer ripstop nylon or equally downproof satins and taffetas.

Cotton ticking or unbleached muslins are popular down retainers for pillows and cushions. Slipcovers can be made for both decoration and ease in cleaning.

Once you have chosen your garment or bedding material and prepared it to receive down, measure the amount of down you want in each part of the item and insert it. If it is to be quilted later, simply insert the down and close up the compartment.

Use long stitches; ten to an inch is sufficient.

Dry out the down by tumbling on low heat, or lying flat in the sun, spanking it occasionally to break up the lumps. Then you can spank the down evenly throughout the compartment and make your crossstitches. Don't worry about catching some down in the stitches; it can't be avoided.

If you are an experienced clothing maker, you can try a more complex item that uses tube construction, such as a cold-weather sleeping bag. I've found the best way to insert the down all the way in these tubes is to get a piece of plastic pipe that fits into the tube and then make a plunger to fit inside the pipe. This can be made with a dowel.

Work the damp down into the pipe, then shove it into the tubes with the plunger. You can work it down by shaking the bag. Don't worry about lumps; they will disappear when the down dries and regains its loft.

How to measure down: As an experiment, I recently ran a test on a pillow filled with northern goose down to find a simple way to measure and weigh down. I soaked the pillow in rather hot water, using a mild detergent, then shook the wet down away from one end. After squeezing away the excess water, I removed enough down to fill one cup and presed it firmly in the cup.

When the cup of high-grade down dried, it weighed 1.74 ounces (49.4 grams). Using this as a base, it would take roughly nine cups of wet down to make one pound of dry down.

Keep in mind, though, that this was unused down. Older down, down that is mixed with feathers, or duck down may require from 50 to 100 percent more to provide the insulation of northern goose down. Unless you know the quality of the salvaged down, it is always best to give yourself some leeway by increasing the down.

It isn't scientific, but one method of testing down for its quality is a simple fatigue test. If you sit on a down pillow or cushion for a half-hour, high-quality down springs back rapidly to its full loft. You can do the same thing with a jacket or sleeping bag by watching a half-hour television show while sitting on it.

Eddie Bauer

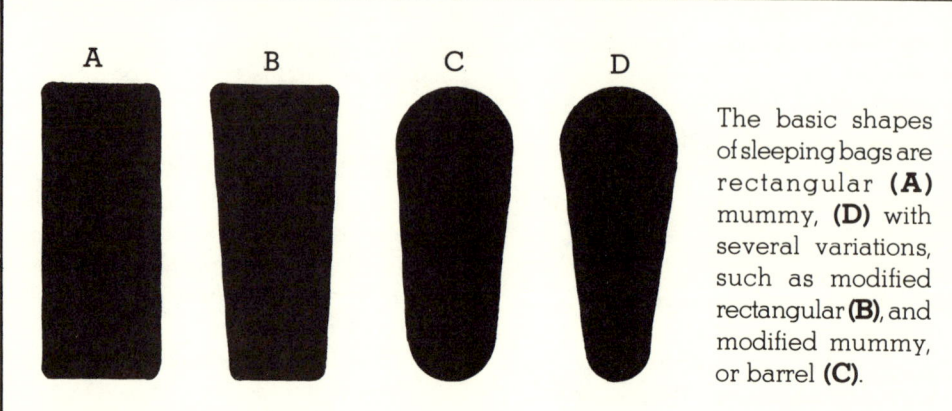

The basic shapes of sleeping bags are rectangular **(A)** mummy, **(D)** with several variations, such as modified rectangular **(B)**, and modified mummy, or barrel **(C)**.

they permit less convection of cold air. They are also more compact and easier to carry. Most mummy bags have an insulated hood with drawstrings around the shoulder area to keep them snug and prevent warm air from escaping. Hoods on the best models are so large that you can cover your entire head, leaving only a small breathing hole. Some hoods in the more expensive bags have a pocket on the ground side in which you can stuff a jacket, sweater, or inflatable pillow for more comfort.

The rectangular bags are usually heavier than the mummy shapes because they use more material. They come in weights suitable to a range of conditions, from warm summer nights to Arctic winter. They have an advantage in that they can be completely unzipped and turned into comforters. Or two can be zipped together and turned into a double bag. They are also ideal for backyard slumber parties and for use in RVs and summer-cabin vacations.

The next consideration is materials and construction. We discussed goose down and synthetics earlier. Since down is much lighter than its competition in synthetic materials, it offers the advantage of lightness.

An average goose-down bag will keep you comfortable in a wide range of temperatures, from ten degrees below zero to fifty degrees above. If the bag becomes too warm in the summer, it can be partially unzipped. For increased warmth, you can wear wool clothing or goose down while inside the bag.

Sleeping bag covers are made of a variety of materials. Down-filled bags must be made of strong and tightly woven fabrics to avoid leaking down through the weave. Hi Count Taffeta or ripstop nylon is used in many goose-down bags because it is virtually leakproof. Another popular material for the fabric that touches your skin is Tri Blend, a blend of polyester, cotton, and nylon that gives the comfort of cotton next to your skin while resisting shrinkage. A third material is Ramar, a rugged poplin blend of 60 percent cotton and 40 percent nylon that is extremely resistant to tears and abrasion and combines the strength of nylon with the comfort of cotton.

The methods of construction are important, too, because they help determine not only the warmth of the bag, but also longevity. The major methods for goose-down bags are as follows:

Sewn-through: The interior and

exterior fabrics are sewn through to form tubes, or channels, to contain the down or other insulation. This technique is common in lightweight bags for use in mild temperatures. It is of limited value for cold-weather camping because the stitching leaves uninsulated areas that allow heat to escape.

Slant-box: Slanted interior walls are created to form baffles that overlap all seams and down channels. The best bags use a stretch fabric in the sidewalls to accommodate the stretching and flexing of the body. This method eliminates cold spots and helps distribute the down evenly throughout the bag.

The most common construction methods for synthetic insulations are the following:

Double offset quilt: The seams of two quilted layers of insulation are offset to maintain even lofting and prevent cold spots. This method is used in bags designed for moderate weather.

Sandwich method: An inner layer of insulation is sandwiched between two quilted layers of insulation and attached to the bag at the edges to prevent shifting. This is a common method used for cold-weather bags.

Other factors to consider are those not immediately obvious, such as the following:

Proper stitching: Eight to ten stitches per inch is best. Fewer stitches may snag or loosen; more may cut or weaken the fabric.

Differential cut (in all bags): The outer shell of mummy bags is tailored larger than the inner shell to permit full lofting and allow you to flex your knees and elbows without creating thin cold spots.

Side block baffle (in goose-down bags): A baffle is inserted opposite the zippered side to keep down from shifting from the top to the bottom.

Baffled foot section (in goose-down modified mummy bags): A baffle is placed across the end of the bag to maintain even loft and warmth.

Differential loft (in modified mummy

METHODS OF DOWN CONSTRUCTION

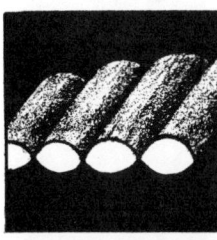
Sewn-through

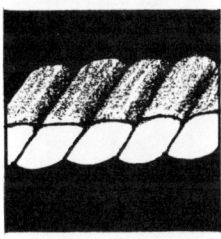
Slant-wall

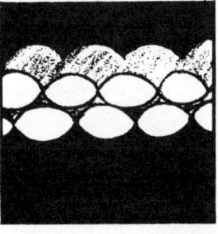
Offset quilt

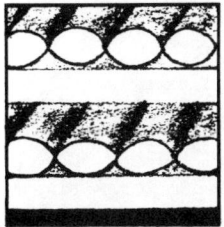
Double offset

bags): Sixty percent of the insulation is on top, where the body weight does not compress it, for better heat retention.

Visible features to watch for include the following:

Sturdy closures: Look for durable zippers, snaps, Velcro fasteners, and draw cords that operate smoothly. Do not settle for second best since these closures are subject to extreme stress. Two-way zippers are preferable because they allow easy and variable ventilation control.

CARE AND STORAGE OF SLEEPING BAGS

When you return from a camping trip, a sleeping bag should be laid out flat to air out and to allow moisture to escape. Then it should be stored loosely to allow the seams and insulation to relax. Do not store your sleeping bag in its stuff bag for prolonged periods because it may cause the insulation to break down and places undue stress on the construction. A good rule of thumb is to use the stuff bags only when transporting the bag.

Goose-down bags can be dry-cleaned or hand-washed. If you have yours dry-cleaned, choose a shop experienced in the care of goose-down products. Check with your outdoor equipment store for recommendations—all Eddie Bauer stores keep a list of shops with experience in cleaning down. A shop that does not have this experience can damage the down.

If you decide to hand-wash, use a detergent designed especially for down, available at most outdoor equipment stores. Wash gently and rinse thoroughly, then tumble dry at the cool setting. It takes a long time to dry a bag or coat thoroughly in an electric dryer.

Synthetic insulation can be machine-washed with a mild detergent. Tumble dry at a cool to warm (not hot) setting.

Do not dry-clean synthetic insulation. The dry-cleaning chemicals will damage the insulation.

PADS

Whenever down or synthetic insulation is pressed flat against the ground, it loses nearly all its insulation value. Down is more susceptible to this than synthetic insulation. Before the development of closed-cell plastic foam pads, campers had to rely on evergreen boughs, leaves, extra clothing, or more covers. The foam pads are seldom more than an inch thick and are cut to the same length as your sleeping bag. The most popular material is Ensolite, and the closed-cell construction forms a barrier between your body and the ground. It insulates so well that it enables you to sleep in comfort on a glacier or lake ice, provided you don't toss and turn and roll off the pad.

The closed-cell foam pads are lightweight and easily roll into a tight package, and they can be rolled separately or wrapped around your sleeping bag, giving it some extra protection if it is being carried through brush or across boulders. The pads can also be used for sitting on beaches and can be stacked in one end of the tent during the day.

Another alternative is the nylon-covered pad made of polyurethane foam. The washable cover protects the pad from snagging and tearing, and weighs only two pounds for one and a half inches of insulation and padding.

Pads come in a variety of thicknesses ranging from one-half inch to two inches. If you plan to combine backpacking with family camping, you will probably want a thinner pad to hold down both weight and bulk. A three-eighths-inch pad is about the minimum that will give you both comfort and protection from the cold.

AIR MATTRESSES

Most campers can be divided into two opposing attitudes toward air mattresses: those who love them and those who detest them. The detesters are frequently campers who tried air mattresses once or twice several years ago when they were as susceptible to tiny holes as a pin cushion. Or, if they didn't deflate halfway through the night, the mattresses created cold spots on the sleeper's back because the mattress did not offer insulation against the ground, or cold air was permitted to move around inside it.

Other campers consider an air mattress almost as important as a comfortable sleeping bag. These mattresses are especially useful in converting first-time campers to your form of recreation: a first-timer who doesn't sleep well is not going to be eager to go out again.

One alternative is to combine the air mattress with a closed-cell foam pad, provided you can get them to stay together through the night. They are also useful when combined with a camping cot, especially those that are already insulated.

A boon to campers' bedrooms is the relatively new self-inflating mattress. These are made of a plastic foam and are covered with a waterproof nylon. They are inflated simply by removing the cap and letting air rush in when the foam springs back to its normal shape. Since the covers are waterproof and airtight, the air inside is heated by your body and does not escape. The mattress rolls into a tight bundle for carrying and storing. A half-length pad weighs one and a half pounds, and a full-length (seventy-two inches) pad weighs two pounds.

COTS

An alternative to sleeping on the ground with the foam pads beneath you—admittedly, they are rather firm—are the insulated camping cots recently made available. Until these cots were developed, few campers used cots during cold weather because they were so cold. But now it is possible to sleep off the ground and still be warm.

Camping cots are popular in fixed campsites, such as hunting and fishing, and when space and weight are no particular problem. They are also useful back home for guests or for drowsing in the backyard. New compact models on the market fold into a small bag not much larger than a sleeping bag.

HAMMOCKS

These sling beds, invented by the Mayan Indians of the Yucatán Peninsula, are sometimes used by campers for fun or as a pleasant accessory. They are lightweight and stow away in a pack or cargo bag. They are easily strung between two trees neatly, and are a very nice place to spend those hot, lazy afternoons while you wait for the sun to drop behind the trees and the fish to start rising again. Few people use them as a bed, although the Mayas have for centuries.

CAMPER'S SECRET

Small pumps, either lightweight foot pumps or the old-fashioned bicycle tire pumps, are still the best for inflating air mattresses because you invariably get some moisture in the mattress when you inflate it with lung power. When you inflate a mattress with a pump, you won't hyperventilate and reel around the campground seeing stars at midday. A sprinkle of talcum powder or cornstarch helps keep the insides dry so that the compartments won't adhere to each other.

CHAPTER 6

THE CAMPING WARDROBE

Preparing for bad weather on a hot day is something like going to the supermarket just after a big meal—not a very appetizing task. When you leave for a camping trip with the sun beating on your back as you load the car, it is difficult to think in terms of warm jackets and sweaters and rain gear. The children will grumble that they don't want to carry those hot old sweaters and they simply hate those stuffy old raincoats Mother always makes them take along.

While there are many campers who seem to survive well in jeans, T-shirts, and tennis shoes, it is courting extreme discomfort, if not danger, to go unprepared for the worst. It is far better to follow the old adage: always expect the worst and you'll never be disappointed.

In warm, clear weather, jeans and T-shirts are perfectly acceptable for campground wear. But come evening or early morning, a goose-down vest or a heavy shirt or wool sweater will be welcome, if not necessary. A sudden shower can drop the temperature immediately, and if you are camping in the high country, that temperature drop can be sufficient to cause primary hypothermia, which is marked by uncontrollable shivering. (See CHAPTER 10 for more information.) Obviously for cold-weather camping your clothing considerations will be different. (See CHAPTER 11.)

Dressing for camping is essentially common sense. Yet, because camping seems removed from normal circumstances, many people are baffled by it and tend to forget that it isn't like home, where you can step inside the house

The layering system of dressing is essential for outdoor comfort and safety. Start with wool socks and warm underwear; add a pullover sweater over your shirt, a parka and hat, and a poncho or rainsuit for rainy weather.

when the rain comes or turn on the furnace when the chill arrives with darkness.

In general, the best insulation for clothing is natural fibers. Cotton is best for hot weather because it gives protection from the sun while allowing air to flow through the fabric to your skin. Wool shirts and pants are the best for warmth and also provide the most protection from moisture in case of rain or snow. Wool can absorb more moisture than any other fabric and still offer warmth. Goose down is the best insulation of all for jackets and sleeping bags.

However, synthetic fabrics have their own values and uses that make them indispensable for camping clothing. The best rain gear is made of nylon coated with a waterproofing agent because raincoats, rain pants, and ponchos made of this material are very lightweight and fold or roll into very small packages when not in use. Coverings for sleeping bags are made of various petrochemicals, as are tents and shell parkas that slip over sweaters and coats as windbreakers.

Many sleeping bags and lightweight parkas and coats are insulated with synthetic materials. These are recommended for beginning campers, particularly those with young children, because they cost less and are simple to keep clean. Also, the bit of extra bulk that synthetics produce is of less consequence than on backpacking or bicycling trips where weight and bulk are primary considerations.

A word of warning to fastidious parents: you may as well forget about keeping the children looking as though they are ready to sing in a school concert. Camping is a vacation from clean clothes and bodies. If your child is a magnet for dirt like Pigpen in the Peanuts comic strip—and most children are—you had best forget about your rules of cleanliness except at mealtime and bedtime. Getting grubby is part of the fun of a vacation for children, and getting to wear the same grubby jeans two days in a row is positively wonderful to active youngsters.

Beginning from the ground up, these are the basic articles of clothing suggested for a weekend camping trip. An additional change of outerwear and two more changes of underwear will usually get you through a week with perhaps one laundry session.

FOOTWEAR

The old joke that the big toe is the thermostat of the body has an element of truth. If your feet are cold or causing pain, you are far from a happy camper. Footwear for outdoor use has become as specialized as all other types of clothing, and it is possible to have a closet virtually filled with shoes and boots for every conceivable use.

For the purposes of this chapter, we will concentrate on footwear suitable for summer camping. CHAPTER 11 deals with footwear and other clothing needed for stretching the camping season throughout the year.

A cardinal rule of camping is never to wear a pair of shoes or boots that haven't been worn for at least several hours before the trip. Only by wearing them will you break them in and find potential blister spots or other problems.

Most campers can get along very well with no more than two pairs of shoes or boots, usually one pair that is suitable for day hikes and normal camping activities, and a second pair for wet-weather use. Since the majority of camping areas are near streams, lakes, or an ocean, a

lightweight pair of waterproof boots will permit you to wade streams or explore the shores without getting wet.

In recent years the National Park Service has begun outfitting some rangers with lightweight walking or running shoes to protect the trails and meadows from the deep indentations made by the lug-soled boots commonly known as waffle-stompers. The old environmental slogan "Take only photographs, leave only footprints" is being rewritten to avoid even footprints if possible, especially in areas of heavy use.

Blends of wool and nylon, or cotton or pure wool socks are preferred by most campers. All are ultimately more comfortable against the skin than synthetics. The first choice is usually blends of wool or cotton with nylon or polyester, which are not only cheaper but also help the socks retain their shape through repeated washings.

Moccasins are very popular among campers who stay at one fixed camp. They are a boon for getting in and out of tents (where you should never wear shoes to avoid damaging the fabric) because a tap of the toe against the heel of the other foot removes them easily.

For those cool evenings and mornings, particularly in the high country where the temperature might drop to freezing at night, goose down-filled booties are excellent for around the camp. Many models have leather or synthetic soles with down insulation from the top of the foot to the ankle. These are easy to keep clean and can be worn in the sleeping bag for a bit of extra comfort on cold nights.

An alternative is slipper socks, with the same sole but heavy wool sock material over the ankle.

In most situations, however, a pair of comfortable shoes good for walking along

THE CAMPING WARDROBE

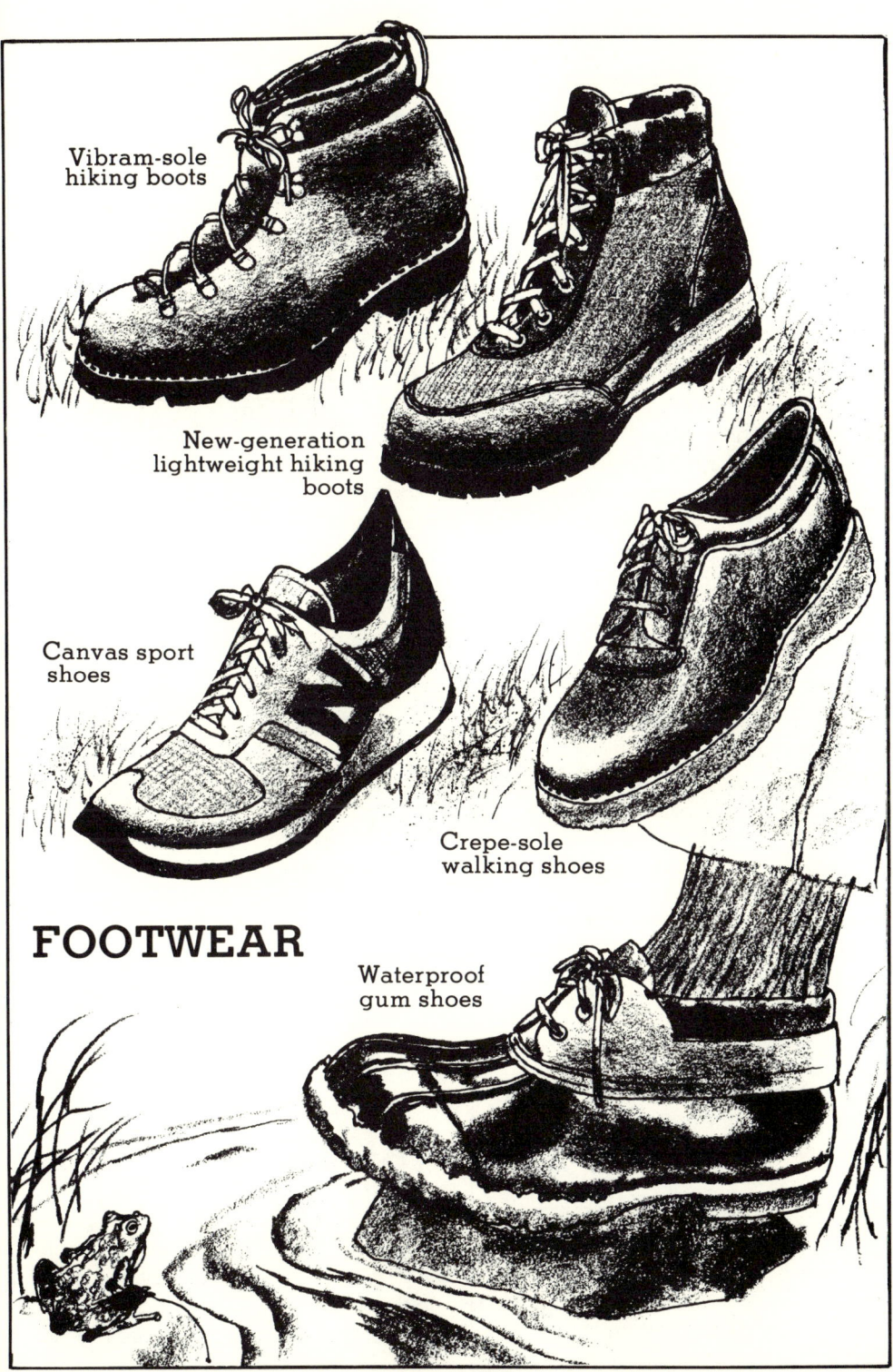

Vibram-sole hiking boots

New-generation lightweight hiking boots

Canvas sport shoes

Crepe-sole walking shoes

FOOTWEAR

Waterproof gum shoes

established trails, plus a lightweight pair of waterproof boots, will be sufficient. Extra warmth can be gained by slipping on a pair of wool socks over your regular socks.

PANTS

Jeans and shorts are suitable for most outings. Try to keep one dry pair always on hand in case of unexpected dunkings or showers.

If you are camping in areas where chilly nights are common, a pair of woolen pants can be taken along in place of one of your pairs of jeans. They are good for evening wear and in case of rain since they continue to provide some warmth even when wet.

SHIRTS

Sturdy washable shirts, preferably long-sleeved for protection against night chill, sun, and insects, are suitable. As with pants, one shirt should be woolen to ward off the chill and damp weather. T-shirts are fine for the heat of the day, but they should be considered a foundation from which you can build layers of clothing, from shirts to jackets or rainwear.

UNDERGARMENTS AND NIGHT WEAR

Although some campers have been known to buy military-style khaki undergarments, not because they are superior garments but because they don't show dirt as readily as white ones, this is still playing sleight-of-hand tricks with yourself. Comfort is the primary consideration. Be sure they don't bind or chafe.

The only other suggestion is that you take along either wool-blend long johns for colder weather or goose down-insulated sleeping garments for the really cold nights and mornings.

Some campers prefer sleeping in their clothing, but this isn't recommended. It is better to take along a pair of lightweight pajamas. If you must sleep in your clothes, make sure they are dry and clean, or at least free of caked dirt and sand to avoid damaging the sleeping bag.

JACKETS AND PARKAS

Under normal summer conditions, a lightweight jacket is usually sufficient for car camping. A school jacket or a ski jacket is fine, as is a lightweight down jacket.

CAMPER'S SECRET

When moving from campground to campground, take advantage of the car's motion by letting it help with the laundry. Place the dirty clothes in a sealed container with detergent and water and wedge it between other containers. The car's motion will act as an agitator. When you arrive at the new camp, just rinse your wash and hang it up to dry.

THE CAMPING WARDROBE

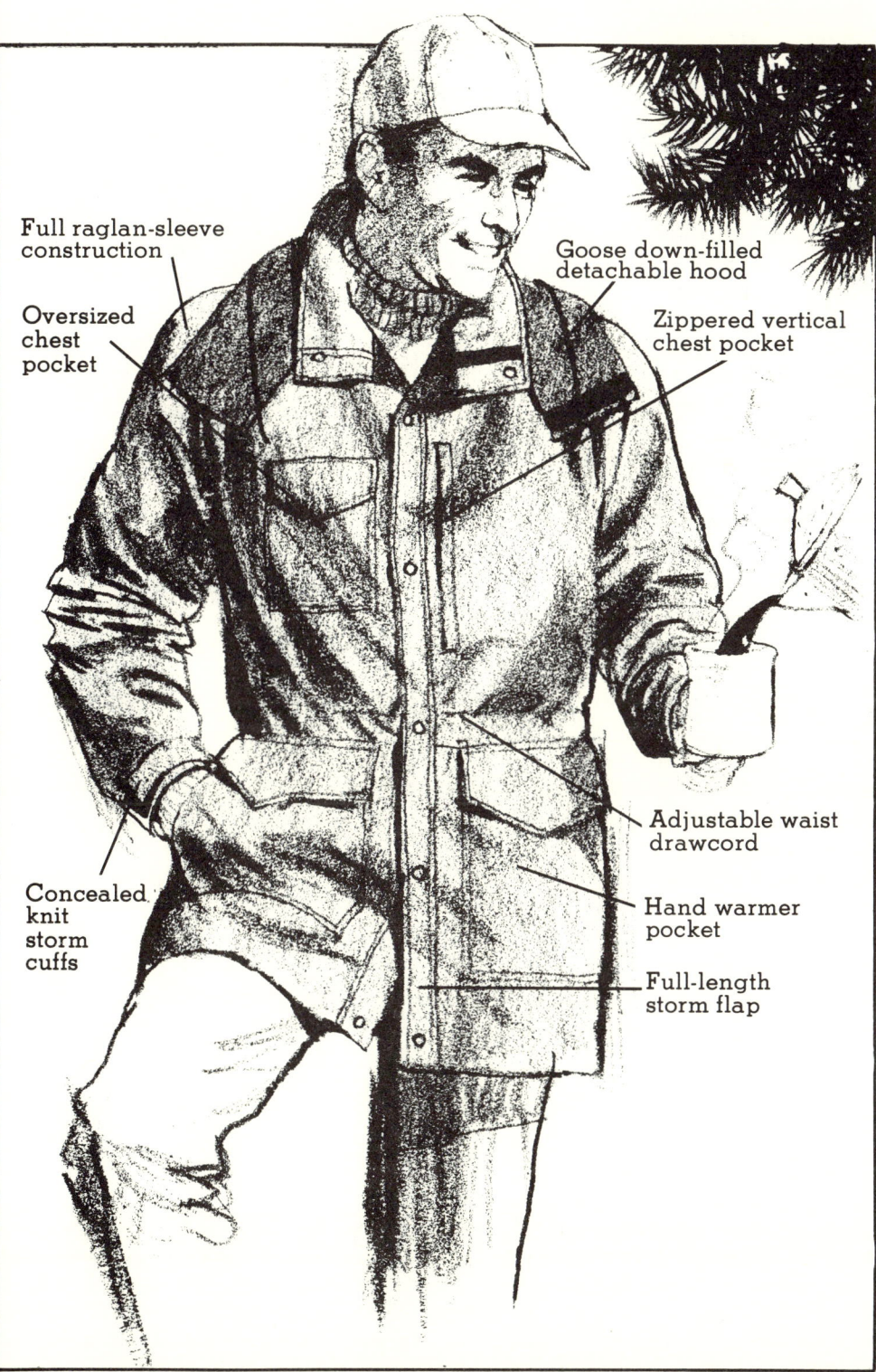

An alternative is to dress strictly on the layer system and, instead of carrying a bulky jacket, use a nylon shell that is called either a wind parka or simply a shell. This is extremely lightweight and will fold into its own pocket to make a wallet-sized package. This, plus a warm shirt and a wool sweater, offers both warmth and protection from light rain.

Parkas come in a variety of styles and are popular casual clothing in many parts of the country. Some are lined with a satinlike nylon or, for colder conditions, with a flannel lining.

Among the many advantages of parkas over jackets are that parkas drop below the hips and keep your waist and kidney area warm, and that they are usually constructed with a hood.

GLOVES

Each member of your group should have a pair of gloves or mittens, even if you're camping in August. Not only do they keep your hands warm on chilly nights, but they are also useful in handling hot pots and sooty grills and pots, and they protect your hands while performing campground chores. In extremely cold weather, mittens are far superior to gloves, especially those with removable wool liners.

HEAD GEAR

You should have some kind of head gear along even though you may never wear a cap or hat in town. More heat is lost through your scalp than any other part of the body, and few things feel more welcome in chilly or wet weather than a

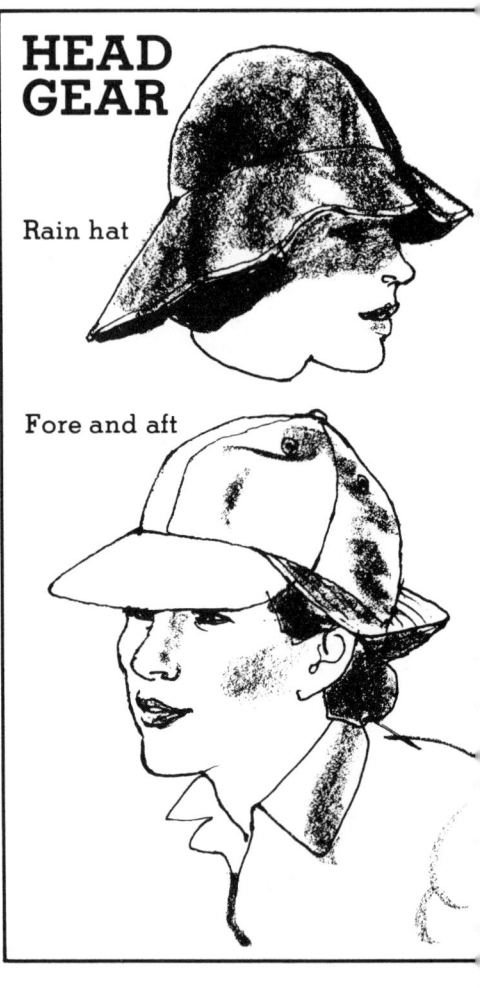

HEAD GEAR

Rain hat

Fore and aft

warm wool watch cap or ski cap. Although nearly all parkas have hoods, few of those designed for summer use have much insulation. Thus, a watch cap or similar cap beneath the hood makes a big difference.

RAIN GEAR

Rain gear comes in a variety of coated fabrics and designs, and each has its own avid supporters. Until only recently most campers said that none of it was

really good; some kinds just weren't as bad as the rest. This was because rain coats and pants were almost invariably stiff and heavy. Since the fabric kept moisture out, it also kept moisture from the body in. Before long, a camper was as damp from perspiration as he or she would have been from standing outside in the rain.

The perfect rain gear still hasn't been invented, but a new fabric on the market is getting closer and closer. The first generation of rain gear was the famed "oilskins" worn by seamen. These were much too stiff and heavy for campers, however, so when nylon came onto the market around World War II, manufacturers seized upon this lightweight fabric as an alternative. By impregnating nylon with waterproof solutions, they gained a very lightweight fabric that was also much more pliable.

These fabrics still had their disadvantages, and it wasn't until the 1970s that a third generation of waterproof fabric was developed. Known in the trade as PTFE fabrics, their trade names include Goretex® and Klimate®.

This fabric was a miracle for outdoors lovers. Although it is porous and

CAMPER'S SECRET

A poncho is one of your most versatile articles of clothing. It can be used as an emergency rain fly, as a sunshade, as a windbreak, as a ground cover for your tent or a seating area for a picnic, as a privacy curtain in big tents or around the latrine. Most come with grommets on the corners and sometimes scattered along the edges so they can be laced together. If they don't have grommets when you buy them, grommet kits are inexpensive and easy to use. Another use for a poncho is as a temporary tent to mark a campsite in a busy campground. Stretch out the poncho like a pup tent, then dash back to the ranger station to pay for the campsite. And two of them strung over a line between two trees makes a good fair-weather tent or a place for you to stretch out away from the crowd for an afternoon nap.

permits perspiration vapor to flow through the fabric and disappear, it is totally waterproof and will not permit moisture to come through to your clothing and body. This microporous film is sandwiched between two layers of outer and inner fabric, such as cotton or nylon (usually the latter).

The first batches of the material had problems. It had to be kept absolutely clean to function; even a fingerprint could plug up the micro opening. And it had to be cleaned frequently with denatured alcohol to keep it efficient. But new laminates have been developed that can be washed using warm water and mild detergents.

Since the introduction of PTFE fabrics on the market, field testing has gone on constantly and modifications continue to be made. But no other product on the market approaches it for practicality.

Following is a rundown of basic rain gear designs.

Coat: Many campers still prefer a simple waterproof coat that can be matched with pants. The coat can be zippered, buttoned, snapped, or pulled over the head. Most designs have a hood attached, and most have some method of keeping the cuffs snug, using Velcro strips, elastic, or snaps.

Ponchos: These are popular with campers everywhere, as they can be slipped on over a backpack and they allow lots of room inside for movement and ventilation to evaporate perspiration. They can also double as a ground cloth for tents, as a tarp for a tent vestibule, or as a shelter over the cooking area. In a sense, ponchos are little more than a tarp with a head hole and hood (which can be tied off to form a waterproof tarp) and snaps along each side.

Their disadvantages are obvious. Ponchos tend to flap in the breeze, even when tied securely to the body with a belt or piece of rope. They catch on underbrush, make a lot of noise, and can be awkward around the campfire or stove.

Cagoule: These are pullover parkas that reach down to or below the knees. Cagoules are almost portable tents because most are cut spaciously enough for you to sit down in them, pull them over your knees and feet, and let the rain fall. The bottom can be wrapped upward and tied around the waist for more freedom while walking. Because they become personal tents, so to speak, they are popular with mountaineers and other people in harsh climates. They are an extension of the parka design used by Eskimos for centuries.

CAMPER'S SECRET

Since up to half of your body heat is lost from the head and neck, where the blood flows closest to the surface, a wool turtleneck sweater and a wool cap will conserve most of that heat.

PART III
AT THE CAMPSITE

CHAPTER 7

SETTING UP CAMP

FINDING THE CAMPSITE

If you plan to stay at one campsite longer than an overnight stop, you should try to shop for it as carefully as you would a hotel or resort. This will, in effect, be your neighborhood for the duration of your visit, and you will want it to be as pleasant as possible.

First, do some research before leaving on the trip. Word of mouth can be trusted only so far, and brochures usually accentuate the positive and ignore the negative aspects. Government campgrounds are usually described reasonably accurately, but these information sheets or little tent symbols on maps won't tell you if you will be camped next to a swampy area, if this particular campground is a popular hangout for teenagers with their hot cars, or if the water tastes as though it had been drained from the radiator of an elderly tractor.

These are the exceptions to the rule, of course, but such things do happen if you're not careful in your campsite selection. There are too many campgrounds available throughout America to have to put up with nuisances on a regular basis.

Federal and state campgrounds are usually the best maintained, especially the more elaborate campgrounds in national forests or those in large parks. Some of the nicest and most peaceful ones are tucked away several miles from the nearest busy highway or town. The

farther you go from the rest of the population, the more privacy you usually can expect.

If you are going out for the first time, or going into an area new to you, it is best to load up on brochures and maps well in advance of the trip. Write the federal and state agencies listed in PART IV (RESOURCES) and get as much information as possible. Check with your local outdoor store for suggestions and other publications on camping in the area you will visit.

Most dedicated campers soon accumulate a bulging file of maps, books, brochures, and other material on campgrounds. You can store this literature in a filing system of your own design. An inexpensive cloth or plastic case will do, as will a briefcase that has seen better days. Take along the literature you'll need for the trip at hand and leave the rest at home.

If you are traveling in an RV, set aside an area near the driver's compartment for maps, guidebooks, and other literature and leave it in the vehicle all the time, extracting only those items needed for each particular trip.

Nearly every major camping area of the country has a selection of booklets and books on the natural history and hiking trails of the area. The National Parks Association connected with the major parks in the system is a good source of guides to the parks and the surrounding area.

A smart rule to follow while searching for the elusive perfect campground is to give yourself plenty of time to explore. If you are leaving on a Friday for a weekend's camping and don't know

SETTING UP CAMP

the area well, find an alternate campground for Friday night on the way, then leave early the following morning to search for a spot to stay Saturday afternoon through Sunday afternoon.

Once you've settled on the campground, you need to stake out the specific site. Ask several experienced campers to describe the perfect campsite, and you'll get a wide variety of responses. It may be easy to visualize the perfect one, but you'll probably never find it. Instead, after trying several, you will soon develop your own list of favorite places and keep returning to them from time to time throughout your camping career. Some campers become attached to a single site and visit it exclusively for their only camping trip of each year.

The basic considerations in choos-

CAMPER'S SECRET

When camping in a desert, never set up camp in a wash, those dry stream beds that look as though water hasn't flowed through them in the past century. Flash floods are common in the desert, and a rainstorm several miles away from you that leaves your area dry can send a flash flood down the wash in which you're camped with no warning. Many campers have been drowned in this manner. Keep to the high ground, always.

ing a good noncommercial campsite include scenery, prevailing weather, privacy, nearby activities, distance from home, and other similar factors. In general, a good campsite should have these characteristics:

- a good source of pure water nearby
- a site high enough to avoid being inundated by a nearby stream
- a site that is level and not subject to flooding in case of rain
- soil that is hard-packed where tents and kitchen will be placed so that tent and awning stakes will hold
- a good source of firewood
- a frequent breeze to help keep the bug population down and to blow away campfire smoke
- sunlight in the morning and evening but shade in the hot midday
- a suitable spot for a latrine, if no public bathrooms are available
- some security against fires
- nearby trees for stringing clotheslines and hanging food to keep away from wildlife

If you are a beginning camper, you might consider making your own check list of campsite requirements, but after a few outings you will instinctively look for the characteristics you require.

Some campers start with a list of requirements, then throw it away after a few trips because they find themselves returning again and again to places that offer few of these amenities; they simply like those more obscure sites because they are close to a waterfall or are high on a shelf of land above a canyon, or for other personal reasons. Sometimes you will find a campsite that is tucked back in a dense grove only a few feet from a well-used trail, giving you the privacy of a patio just off a freeway exit. Campers have been known to become very secretive about their favorite haunts and to share their locations only with family members or close friends—something like a condominium partnership.

Setting up a campsite is largely a matter of personal preference, much like decorating a home. Yet there are basic components of a campsite just as there are of homes: living area, bedrooms, kitchen, and bathrooms.

In most campsites where campfires are permitted, the fire ring is usually in the middle of the site. The natural terrain will dictate, to a great extent, where the other components are placed. You will want your tent or tents erected on level spots to avoid sleeping on a hillside, and preferably at least fifteen feet from the campfire to avoid the danger of sparks landing on the tent. The kitchen should be convenient to the campfire, yet a few

Hang food out of reach of bears and other animals.

steps away so that people can ring the fire for warmth or run from the smoke without blundering into pots and pans. The bathroom will be as far away from the campsite as possible and downwind from the prevailing wind.

In more organized campgrounds, such as most government and privately owned sites, these factors are already taken care of for you. All you need do is park in the assigned space, pitch the tents, and unload the kitchen onto the picnic table.

When you pitch the tents, it is best to have them facing away from the prevailing wind. This helps keep the breezes from whipping the tent door during the night and will prevent rain from being blown in. This is a consideration only, and not critical with most modern tents, which have rain flies that completely cover the entrance. Also, when more than one tent is being used by a family, it gives a sense of security to have tents facing each other.

It is a rare camper who cannot set up a camp without rearranging the "furniture" according to personal tastes. Many campsites have stumps or logs or big blocks of wood suitable for sitting or use as a kitchen table. Part of the fun of setting up a campsite is organizing your temporary home.

A good schedule for setting up camp begins with reaching the campsite well before dark to save much frustrating fumbling in the dark, and to give yourself a wider selection.

On arrival at the site, most campers first set up the tents. Then the sleeping bags are stretched out or hung from nearby trees or over bushes to air out before dark. Then the kitchen is laid out and a campfire started. Keeping in mind that many government campgrounds, particularly national parks and national

BEACH CAMPING

Camping along the seashore and some stretches of the Great Lakes presents different problems for campers to resolve. The major one, of course, is the presence of sand. The other, on seashores, is the tides.

You will need something to replace your tent stakes because they will not hold in sand. The "deadman" weights made of stones or pieces of boards will help, as will snow anchors made for winter camping. Both of these anchors are tied to the tent lines, then buried and stamped down firmly.

Since you will be standing, walking, and sitting in sand, it is going to be everywhere, especially in the bottom of the tent. If you have large pieces of tarp or unused ponchos, these should be stretched beneath the tents and out a few feet to form a patio of sorts. Sand should be swept out of the tents constantly, and everyone should sit down in the entrance of the tent and remove shoes and socks and shake them out before entering the tent. Sand is an excellent corrosive and will wear holes in the floor of a tent rapidly. At the best, it will weaken the floor more in a weekend's use than a month's normal use.

Another potential problem on seashores is the tides. You will see the tideline along the beach, and no matter what the tide table says, you should never take a chance on sleeping below the last high-tide line. Few things are more frightening and dangerous than waking up with saltwater lapping in your tent.

wilderness areas, do not permit campfires except in the heavily organized campgrounds, you should be aware of such restrictions before leaving home. In any event, you should always carry the campstove and ample fuel for the trip. It is best to call ahead and see if firewood is available at the site.

If you are in a more primitive camping area with no firewood available for purchase at the ranger's office, wood gathering is one of the first chores after a site has been selected. Government recreation specialists speak of the "human browse line," which refers to dead branches on trees that people pick off for firewood. You will note in heavily used camping areas that dead branches have been broken off up to a height of six or seven feet. This does not damage the trees, but live trees and saplings should not be cut unless prior permission is obtained from whoever owns or manages the land.

If you find poles lying around a camping area, these should not be burned or damaged, because someone no doubt left them there for tent poles or similar uses. Even though you may not need them, the next camper might be using a tent that requires such poles.

One of the most important rules of outdoor etiquette you should always practice is to leave a supply of wood and kindling behind when you leave a campground. This rule harkens back to the pioneer days when users of wilderness cabins always left a wood supply in case the next occupant might arrive suffering from hypothermia, frostbite, or worse, and an immediate fire was a life-or-death matter. Granted, this seldom happens in lowland camping during the summer months, but good manners require that we leave the campsite in better condition than we found it.

CAMPFIRE BUILDING

In areas where campfires are permitted, everyone in camp should be taught how to build wood fires quickly and simply. As each member of the family becomes more and more adept at getting them started, the conditions should be made more and more difficult so that fire-building becomes easy for everyone.

You will find that children enjoy this combination of learning and play; in fact, fire-starting holds what seems to be a primeval attraction to children. The problem is keeping their enthusiasm under control. It is sometimes difficult to convince children of the importance of fire safety in the outdoors when they

FIRE-BUILDING

Tepee fire

Log-cabin fire

Star fire

have used almost a box of matches trying to get dry wood to burn with no success. Just as it is easier to grow weeds than vegetables, it seems very difficult to start a campfire but simple to start a forest fire.

The one essential to starting a campfire is good tinder, something that will burn easily and provide enough flame and heat to ignite the other components of the fire.

Some of the best tinder comes from birch bark, dry moss, dry pine needles, crushed cedar bark, and very small dry twigs. An alternative when you can't find small tinder is to make dozens of notches in a piece of dry wood, leaving the chips and slivers still attached. This will give you both tinder and fuel in one piece.

The Boy Scout test of starting a fire with one match is excellent discipline for everyone, beginner or seasoned camper, because it forces you to construct the fire properly the first time.

Tepee fires: This is the classic structure for campfires. The tinder goes on first, with space between the twigs or needles or whatever the base is. Then slightly larger sticks are stacked loosely around the center, with a space on the windward side for air, either wind or your breath, to enter.

A common mistake made when building this fire—and all others—is to put too much fuel on too early rather than let it get a good start and gradually add the larger fuel.

When the basic tepee has burned, it will leave a good bed of coals for the remainder of the wood, which can be laid flat.

Log-cabin fire: This is a good fire to use for quick warming rather than simply cooking. As with the tepee design, this one starts with the tinder; then add larger sticks stacked log-cabin or rail-fence style all around it, and finally cap it off with a loosely built lid of firewood. This fire will heat much faster and create a bigger blaze than most other types.

Star fire: This is a flat version of the tepee fire and gives you more control over the fire and the cooking heat it produces. As the sticks burn, you shove them into the fire in amounts sufficient to keep the fire going. One chore for a member of the party is to keep a good supply of tinder and kindling available

CAMPER'S SECRET

Unless you are camping right beside a stream or lake, always keep a bucket of water handy in case of fire. Plastic jugs or buckets that collapse are excellent for this. You might carry one that is marked "Fire."

CAMPFIRE SAFETY

More and more camping areas are either banning campfires entirely or restricting them to designated fireplaces or concrete-ringed firepits. It is illegal in many camping areas to cut down any timber, dead or alive, and consequently more and more campers are having to bring their own wood or charcoal. It is a disappointing situation for most campers, who find it difficult to imagine camping without the smell of woodsmoke and the fascination of staring into a campfire at night.

However, fires are still permitted in many areas and you should never lose your fire building skill or fail to pass it along to your children, if for no other reason than for use in emergencies. Following are some dos and don'ts of campfire building.

- Don't build new fire rings when one is already there.
- Don't build a fire on the edge of a water source. Keep it as much as 200 feet from the water to prevent pollution.
- Don't build a fire next to a tree, near exposed roots, near vegetation, or on anything other than bare ground.
- Don't build a fire with a clean rock as a reflector.

SETTING UP CAMP 105

- Don't burn any garbage other than paper and organic trash or other materials that will burn. By putting paper, metal, glass, and other material in the fire, you're only starting a garbage pit.
- Don't leave the campfire unattended.
- Burn only dead wood that is lying on the ground.
- Keep your fire small.
- Use wood that doesn't create sparks, if possible. Green wood, some driftwood, and woods with lots of pitch such as pine, alder, spruce, and hemlock are notorious for sparks.
- When you're done with the fire, literally drown it, then keep poking around the deluged area for bits of burning material beneath the mud. Smooth over when completed.
- If you must build a fire in an area with lots of humus, carefully cut out chunks of the sod and stack them neatly to one side. Dig down to bare dirt and ring with rocks. Then build the fire. When you're through with the fire, drench the entire area with water, and when you're certain it is completely out, remove the rocks and fill the hole with enough dirt so that you can replace the sod to its former depth. If you're neat, nobody will suspect a fire has been there and you've treated the forest properly.

at all times. It is a good idea to store enough for a fire in a container in the corner of your tent or some other dry place.

Trapper fires: After the fire is built, you need something to put the skillets, pots, and pans on. This fire has logs or rocks on each side of it to serve as racks and to concentrate the heat. It should be laid out with the opening in the direction the wind blows. It can also serve the purpose of drying out logs before they're put on the fire.

Reflectors: To send the heat toward your tent or to keep it away from equipment stacked nearby, you can build a reflector of a variety of materials. One of the best materials for the reflector is aluminum foil, but rocks or stacks of wood will also work.

Trench fire: This is a version of the trapper fire, except that it is in a hole in the ground with the sides serving as a rack for your utensils. These are good fires for sandy beaches and you can line the holes with rocks. After you have made certain the fire is completely out, it can be buried with no trace.

As everyone's skill in fire-building increases, the task can be made more and more difficult—setting time limits, using wet wood, or limiting the campers to one piece of wood for building a fire. These tests of skill not only help fill a long afternoon or evening, but also teach survival skills that will make campers more secure in the outdoors, and better camping companions, too.

Preparing the site: In most camping areas you will be using the same firepit others have used before you. If an established fire ring is available, always use it instead of creating another one since fires damage the soil for a few inches down. Also, if the fire ring is changed, before long the entire campsite will have an ash and charcoal floor.

Be sure all vegetable matter—roots, branches, leaves—is removed from the ring. Fires may be difficult to start, but once they get started on an underground root or wood covered with soil, they will travel along it.

Do not build fires against boulders. Although boulders make excellent reflectors, the burn marks on them will last for years, if not decades.

If you are camping on a sand or gravel bar along a stream or lake, try to burn everything in the fire so that all you have left are ashes, which can be buried and will soon wash away.

Putting out the fire: When you are packing to leave the campsite, permit the fire to burn down before you leave. Completely drench the fire area for several feet around it. Keep pouring water into the soil until the entire area is soaked, and keep testing the surrounding soil for hot stones or underground pieces of wood still smoldering. It is far better to leave a sooty puddle behind you than a smoldering stick or root.

Leave enough tinder and wood behind for the next party to get a fire started. It is not only the courteous thing to do, but also a safety measure in case the next group has someone suffering

---

CAMPER'S SECRET

You can make your own fire-starter briquets by using an empty carton. Fill the depressions with a mixture of melted paraffin, sawdust, or wood chips, then cut them apart into individual briquets, wrap in plastic, and carry for use as a fire-starter.

from hypothermia or, less drastic, they arrive late and need a fire for both heat and light while setting up camp. In any event, it is simply basic outdoor good manners.

In the colder regions of North America, Canada and Alaska especially, it has always been a tradition to leave unoccupied cabins with at least as much firewood as when you arrived, plus good tinder or kindling and a good supply of matches and food. Many lives have been saved through this courtesy, and it should extend to all camping areas.

LATRINES

Next to sleeping outside, outdoor bathroom facilities are a major hurdle for many first-time campers to overcome, particularly those who have no experience with life beyond the city limits. This problem is similar to that of the farmer whose wife repeatedly asked him to build a patio so they could eat outside in good weather. "I spent half my life eating in the house and going outside to the bathroom," he replied. "Now you want me to eat outside and go in the house to the bathroom!"

The camp's bathroom facilities are not an insurmountable problem, of course, and after the initial camping trip, most people are so enthusiastic over the pleasures of camping ("The coffee **does** taste better") that the lack of modern bathroom facilities becomes only a minor inconvenience.

If your campsite is not within an established camping area with running water for showers, bathroom, and laundry, there are a variety of choices for a latrine.

Some campers prefer carrying portable toilets along that can be set up behind a screen of trees or brush, or behind a tarp or poncho. Some models come equipped with chemicals and disposable bags, and some simply have a plastic bag that is disposed of after each use. Most of these require a bit of pioneer carpentry to build a base.

Other models are self-contained and designed for use on boats as well as for camping. The base is built in and can be placed on the ground almost anywhere.

This isn't a factor at organized campgrounds where an outhouse is always part of the equipment, and in the more elaborate campground flush toilets are used. However, a word of warning: unless the outhouses are cleaned frequently, using them can be unpleasant. You may want to carry your own portable toilet if you plan on camping at several places you haven't visited before. Again, call ahead to the managing agency for information on this and other areas of concern. We have come so far since the pioneering days that the vast majority of Americans have never used anything other than indoor plumbing, and the thought of not having it available may come as a shock. Thus, it is important to introduce children to camping early. In the case of adults, it might be best to stay at organized campgrounds first, then gradually ease them into the wilderness in stages.

In more primitive situations, however, most seasoned campers eventually come to depend on a latrine that is dug in the form of a trench and filled in as used. Other choices are what some call "cat toilets," meaning you dig a hole as needed and fill it in after use.

All bathroom facilities should be downwind from the camping area, and should never be within two hundred feet of a stream or lake. Human waste should

LATRINES

Portable toilet

Trench latrine

be buried at least six inches or even a foot deep. A good safety precaution is to leave a mound of dirt over the covered hole so other campers won't accidentally dig in the same spot.

It is best to scout out bathroom areas in different directions from camp, one for males and the other for females.

Be wary of building them near places frequented by insects, and be particularly watchful for hornets that may be nesting nearby.

GARBAGE

The whole subject of garbage can be stated simply: pack it in and pack it out. Unless you are at an established campground with frequent pickup of garbage cans in the area, do not scatter leftover food around the area on the assumption that it is biodegradable. Even though it may be, it won't rot and nourish the soil for some time, and it will turn wildlife nearby into scavengers dependent on campers for food. It will attract some of the less desirable elements of the animal kingdom, such as bears, porcupines, and skunks, who will lurk around the campground as a threat to your food supply. Besides, garbage stinks. You may not throw much away, but the campers before you and after you might feel the same way, and before

SETTING UP CAMP 109

"Cat" toilet

a camping season ends, that campsite can become ringed with garbage.

Some campers dig grease pits to dispose of bacon fat and other types of grease. This should be avoided, too. Grease will burn in the campfire. If you are camping in an area that prohibits campfires, store the grease in a plastic container or a tin can and pack it out with you.

Most camping food comes in plastic or foil packs that can be wadded into small balls that weigh virtually nothing. Take along a nylon sack with a drawstring and stuff all wrappers and containers into the sack, carry it back to the nearest garbage drop, then wash the sack when you return home.

CAMPER'SECRET

Although most campers prefer keeping the toilet paper in the main camping area—if nothing else, its absence serves as a signal to others that the latrine is spoken for—some store it on site in a tin can suspended upside down from a tree branch or stuck in the ground nearby.

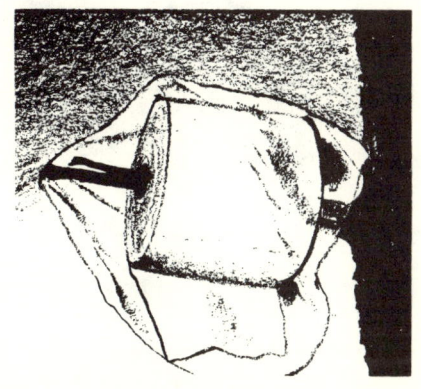

DIVIDING THE LABOR

An odd thing happens to children on camping trips. The lazy one whose idea of a strenuous evening is getting up occasionally to change television channels will often be the most willing worker on a camping trip. While going outside to get a load of firewood for the family fireplace at home is beneath this child's dignity, out in the woods it is often a grand adventure that he or she is unwilling to share with other siblings.

The mechanics of erecting a tent fascinate most children, and learning how to operate a white gas campstove is akin to learning to drive an automobile. Nearly all camping equipment has a toy-like mystique for children, and organiz-

ing work parties and dividing the camp labors usually is not the problem encountered when assigning home chores.

Parents of today who unthinkingly assign camp chores according to gender are going to receive one of the short, pithy lectures on equal rights between the sexes. Girls aren't automatically the cooks and dishwashers, nor are the boys wood gatherers and splitters. Your past experience with the camping companions will dictate how to assign tasks.

Some families work best on the volunteer basis. Something needs to be done, and everyone pitches in to help. While one is mixing the pancake batter, another is mixing breakfast juice and another is replenishing the firewood or getting the campstove going.

In most cases, though, it works best to organize a rotation system before the trip begins and give the children a voice in the proceedings. Who will do the dishes the first day? Who will do the laundry, or will each person be responsible for his or her own laundry? If you are eating freeze-dried or dehydrated food for most of the trip, anyone can prepare it. But who knows how to prepare a mushroom omelet?

Nearly every family has a trader in its midst who can, through a series of complicated maneuvers, spend most of the trip standing around watching while others do all the work. While nothing is wrong with parents letting the children trade jobs occasionally, they should step in whenever they see one child on the verge of taking advantage of another. A feud between children can cast a pall over a camping trip and should be avoided.

Experience at home and on pre-

vious family trips will indicate whether to depend on the volunteer system or to assign tasks. Often it is a mistake to force someone to cook who doesn't know the difference between a fire with moderate heat and one with extreme heat. Each person should, however, know all camp chores, and if you deem it necessary to assign unwanted chores, especially the cooking, let the others decide (within reasonable limits) what they will cook when their turn comes.

Some families prefer making the cook responsible for everything, start to finish, including washing the pots. The wood gatherer knows in advance how much to bring, such as enough for the rest of the evening and kindling and wood to start tomorrow morning's fire. When dividing the camp chores into units, be as specific as possible on what is expected.

Each person should be responsible for his or her own camping equipment; sleeping bags, clothing, eating utensils, and toilet articles. Those using each tent are responsible for erecting it (with perhaps a bit of help from adults), keeping it clean, and stowing it away when the trip ends. This kind of responsibility should begin on the first outing, no matter how young the child is. Most children secretly enjoy responsibility that they can handle. It gives them a sense of importance, and something to do while getting accustomed to unfamiliar surroundings.

In all cases, safety should be the first consideration. Teach children how to handle potentially dangerous equipment, such as campstoves and gasoline lanterns, and how to build campfires. But always supervise them, and do not let them perform these tasks unless one of the parents is nearby.

Only through experience will you learn to organize and run a campsite to suit your own needs and tastes, and until you become seasoned in this form of recreation you most likely will take many more things than you will use or need during the trip.

Unless you are an organizational genius, and your children are compulsively neat, you had best assume that the first outing or two is going to be chaotic at times. Someone will misplace a drinking cup or leave it beside a stream on a day hike. The reader of the family will fall asleep while reading a Nancy Drew mystery by flashlight, and you won't have an extra set of batteries. A field mouse or porcupine will gnaw holes in all of the hot chocolate mix that somebody forgot to put with the other food suspended from a line high in a tree. The family glutton will eat three days' supply of peanut butter at one sitting when nobody else is watching.

Nearly every trip, no matter how many you've been on, has its little catastrophes. But the more you camp, the fewer and more minor they become. In most cases, they become part of a family's legacy of "remember that time" stories that are told for generations.

CAMPER'S SECRET

Beware the birds. In many areas you'll have to watch out for certain birds that can be absolute pests around campgrounds. Crows, or ravens, can be among the worst because they are as intelligent as they are mean. They delight in stealing your food, and can swoop down and fly away with a piece of bread or almost any food they can grasp. They also love shiny objects and will steal them for the fun of it. One camper told of a flock of crows that tore a styrofoam ice chest to shreds because they know that was where the food was kept. Magpies, "camp robbers" and a few other species of fearless birds can be interesting to watch but infuriating come mealtime.

CHAPTER 8

THE OUTDOOR KITCHEN

A happy crew is a well-fed crew. A pot of coffee on the back of the grill . . . hot water for tea or hot chocolate . . . bacon or sausage frying in the morning . . . drink mixes that quench your noon thirst while giving a slight puckering sensation . . . a dutch oven filled with vegetables and meat that is ready to eat just before sundown. . .

A family that knows it will have tasty food on a camping trip will look forward to each outing, and you will hear much less muttering about missing favorite television shows. Poor planning is the only reason modern campers come home grumbling about the food, and often the best compliment paid to the campfire cook is silence on the subject.

Thanks to refinements in freeze-dried and dehydrated food technology, grocery shelves are laden with your choice of food that is as tasty when cooked as it is light in weight. Consequently, those "golden" days of camping are gone when novices wondered if they were expected to supplement their diets by grazing in nearby pastures.

It isn't necessary to pack up your microwave oven and food processor to have well-balanced and tasty meals. An easy way to test that statement is to go through the shelves at your grocery store and see how many items you normally buy that are prepared simply by dropping them into boiling water. You'll find that garden-fresh vegetables are about the only kind of food that doesn't fit into this ultralight—and long-lasting—category.

The major challenge to campfire

cooks, then, is to make life as simple for themselves through the shelves possible. Some campers may genuinely enjoy cooking as a hobby, but not at the expense of enjoying those activities only camping offers.

Family camping trips should be vacations for everyone in the family, especially the cook. By the same token, a group that has delicious and nutritious meals is going to have more tolerance for bad weather, too much wind, or too much heat than a group that sacrifices good food in favor of other considerations. Nor is a camping trip an appropriate time to continue the family struggle to make the youngsters eat their

---

CAMPER'S SECRET

Take along familiar food the first time. If you're going on a weekend trip, prepare at least one good meal—perhaps even catering to the children's preferences so they won't have a memory of cold mush for breakfast.

---

broccoli or spinach; vacations are a time to avoid these gastronomic debates.

Everyone should be involved in the food preparation and cleanup afterward. You will know from past experience if your family can assign a different cook each day, or if it works best to give each member a specific job to perform during the trip. Obviously, each member should be responsible for his or her own tableware, and if one has a history of being careless about matters of hygiene, a parent should hold frequent inspections of cups, plates, and eating utensils to cut the risk of food poisoning due to laziness.

The best rule to follow in food preparation is to keep it as simple as possible. Grocery shelves are laden with excellent foods that require only the addition of water to reconstitute into delicious courses. Food manufacturers have become very adept at producing these for the so-called salt and pepper cooks, a term that aptly describes what much camp cooking is.

COOKING IN THE OUTDOORS

Although campfire cooking is the traditional way, more and more campers are depending almost entirely on stoves because they are more dependable, you have more control over them, and many areas do not permit campfires, anyway. Your camping stove can range from the old standard Coleman two-burner white gas stove to the more recent pack stoves that weigh only ounces. Butane and propane stoves are very popular, too, and come in one- and two-burner models. None of these is excessive in weight, and all are relatively easy to keep clean.

The major problems with cooking over open fires are that you can't control the heat as simply and quickly as with a stove, and you always have the problem of soot on the bottom of your pots and pans. This can be partly alleviated by coating the pots and pans with a thick application of soap or wrapping them in aluminum foil, but stoves remain much simpler to use.

Also, the governmental agencies that administer many of the camping areas are stressing stoves more and more because they are more environmentally responsible ways of cooking.

THE OUTDOOR KITCHEN 117

Stoves: Camping stoves come in three major fuel categories: white gas, propane and butane, and kerosene. Each obviously has its advantages and disadvantages, and each has its own band of loyal users.

White gas stoves are probably the most common in America because the fuel is readily available everywhere in the United States, and the fuel can be either standard white gas or commercially marketed naphtha such as Coleman and Blazo fuels.

These stoves come in two major types: those that must be preheated by pouring a bit of alcohol or gas on the burner, and those with pressure pumps that force fuel into the burners. Under normal summer conditions, these stoves present no problem, but in cold weather the preheating can be a trying and time-consuming chore. This preheating is accomplished by pouring a small amount of fuel or alcohol onto the burner and igniting it with a match. After a minute or so the burner will be hot enough to turn the liquid fuel into a vapor.

The butane and propane stoves have gained popularity for warm-weather camping because no priming is necessary and the fuel comes in cartridges, making it unnecessary to transfer fuel from container to stove. These stoves burn hot and quiet and are by far the simplest to operate.

PURIFYING WATER

Unless your camp is served with treated municipal water, never take chances with its purity. There are several methods of purifying water and you can select the one that is simplest for your type of camping.

Boiling: You should boil it for at least five minutes, longer if it is convenient. Let the water cool and the sediment settle to the bottom, then pour off the pure water and discard the sediment. To remove the bland taste caused by the boiling, aerate by pouring back and forth between two containers.

Purification chemicals: Several chemicals are available to kill bacteria in the water by letting it dissolve and sit for the specified length of time. New products are being introduced on the market all the time, including the so-called water stick, which is a cigar-shaped drinking straw filled with charcoal and chemicals that will purify up to 300 gallons of water. With it you can suck water directly out of a stream, knowing it is pure when it reaches your lips.

CAMPER'S SECRET

A new cookie sheet can be used as a good fire reflector to bake biscuits or pies.

Their disadvantages are the bulk of the cartridges, which must be carried out with you and discarded. And butane, the most popular, will not vaporize at freezing level. If you are going to use it in cold weather, you must literally sleep with it during the night to keep it warm. Propane is not so severely affected by cold weather.

A disposable butane cartridge will provide up to three hours of burning. Propane is a much higher-pressure fuel, and the containers must be heavier to withstand the pressure. Both are sus-

THE OUTDOOR KITCHEN 119

STOVES

White gas Coleman stove

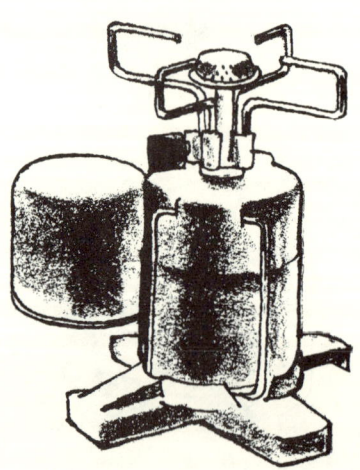

Butane Bleuet stove

ceptible to overpressurizing when left in the hot sun, and safety valves sometimes open to relieve the pressure.

An advantage for car camping and similar activities is that many accessories have been developed for butane and propane, such as small lanterns hardly larger than a flashlight that fit atop butane cartridges, stoves that nest for carrying into a shape not much larger than a saucer, and small heaters.

White gas stoves, as noted earlier, are the most common and versatile of the choices. White gas is used on the one- and two-burner campground stoves and the tiny stoves that nest into cooking pots. Since it is volatile and produces carbon monoxide, white gas must

CAMPER'S SECRET

Fill bottle caps with melted paraffin and place a short length of cotton string inside for a wick. These can be used either as firestarters or for about fifteen minutes as a candle.

be handled carefully, especially in enclosed spaces such as tents. You should make it a rule never to use any stove inside a tent unless the tent is equipped with a zipout cook hole so the stove is on the ground or snow. Tents are treated with fire-retardant chemicals, but most are built of petrolem-based products and **will** burn.

Another disadvantage is the necessity of priming the stove before it will burn. Some models have pumps to create pressure in the fuel tank. Others, which are popular with backpackers and climbers, are the models that must be preheated by pouring a small amount of fuel or alcohol onto the burner. Most

THE OUTDOOR KITCHEN 121

White gas Optimus backpacking stove

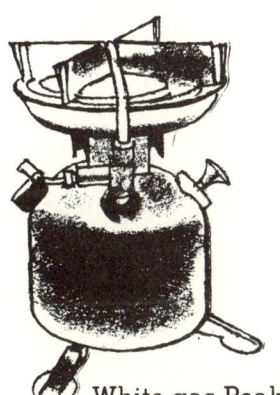

White gas Peak I backpacking stove

Two-piece MSR white gas stove

campers prefer carrying an eyedropper to avoid spilling fuel while pouring small amounts on the burner for preheating, and some prefer carrying a small vial of alcohol because it is less volatile than white gas.

Kerosene is the least popular fuel in America because it creates an unpleasant odor, is susceptible to smoking and staining pots and pans, and is the least volatile of the fuels, meaning it is the most difficult to coax into flame. However, it is worth noting that kerosene comes closer than any other stove fuel to being universal. It is available virtually everywhere in the world, whereas white gas is essentially a North American fuel and propane and butane cartridges are almost equally difficult to find. But kerosene is available on every continent of the world.

Alcohol is almost equally unpopular because of its low heat.

Heat: If you are going to camp in extreme cold—high elevations or stretching your camping season into the winter months—a tent heater will make your mornings and evenings more pleasant, although they add considerably to your weight and bulk and are not suitable unless you camp near your vehicle or haul everything into camp on pack animals.

Most heaters burn some form of gasoline, mostly white gas, which means they must always have ventilation to aviod the danger of suffocation. They are of more use in the heavier, nonporous tents than lightweight models that "breathe" well. Their usefulness in the latter models is limited because heat, like moisture, dissipates rapidly.

Another useful source of heat is hand warmers, which are only slightly larger than cigarette packs and provide heat for around an hour without flame.

Many winter fishermen and fall hunters consider them an essential part of their equipment because one in each pocket keeps hands from getting cold and stiff.

Generally speaking, however, if you have the proper gloves or mittens, you won't need them with you. But they are a good addition to camping equipment to warm cold or wet hands.

> **CAMPER'S SECRET**
>
> Plastic film cans make good, lightweight containers for salt, pepper, herbs, and spices. While traveling, keep the original snap-on lids with labels on the cans. Most camping supply stores now stock shaker lids that will fit over the cans.

CAMPSTOVE SAFETY

- Never fill a stove inside the tent or near an open flame.
- Keep fuel cans away from all sources of heat and in the shade.
- Never fill a stove inside the tent or near an open flame.
- Avoid cooking or heating with the stove valve wide open. It overheats the stove.
- Don't fill a stove while it is still hot.
- Don't carry fuel in plastic containers that can be punctured.
- Memorize the stove manufacturer's instructions and carry them with you.
- If you smell fuel—gasoline or natural gas—before lighting the stove don't light it until you have found the source.

UTENSILS

Since more and more camping areas—particularly national parks—prohibit campfires, it is best to assume you will have to carry your own stove. If you do camp in a place that has a good supply of firewood and you can build a campfire, you will probably still want to use your campstove so that your pots and pans stay clean and use the campfire for hot dogs, marshmallows, and evening cheer.

Nesting pots made of aluminum have long been standard camping utensils because of their light weight. However, recently campers—and even backpackers, who must fret over every additional ounce of weight—have been won over to stainless steel. Aluminum dents more easily, is more difficult to clean than stainless steel, and has a tendency to "shed" its black residue on your hands and towels. Although this microscopic residue is harmless and seldom alters the tase of food, it still bothers some people. Stainless steel is a bit heavier, true, but it is more durable and easier to clean and since it is always shiny, it looks cleaner.

In selecting your cooking gear and organizing menus, simplicity is the key word. Discipline yourself to use as few pots and pans as possible so that packing, unpacking, and camp tending will be less complicated. Some campers have learned to prepare meals for the whole party using only two modest-sized pots and a one-burner stove.

One essential that never should be overlooked is at least one high-quality cooler, preferably with a hard shell so it will take the normal wear and tear of camping. It is one of the most versatile of

THE OUTDOOR KITCHEN 123

Old-fashioned camping cook gear was heavy, often made of cast iron, but today's equipment, shown below, weighs only ounces and nests together to save space.

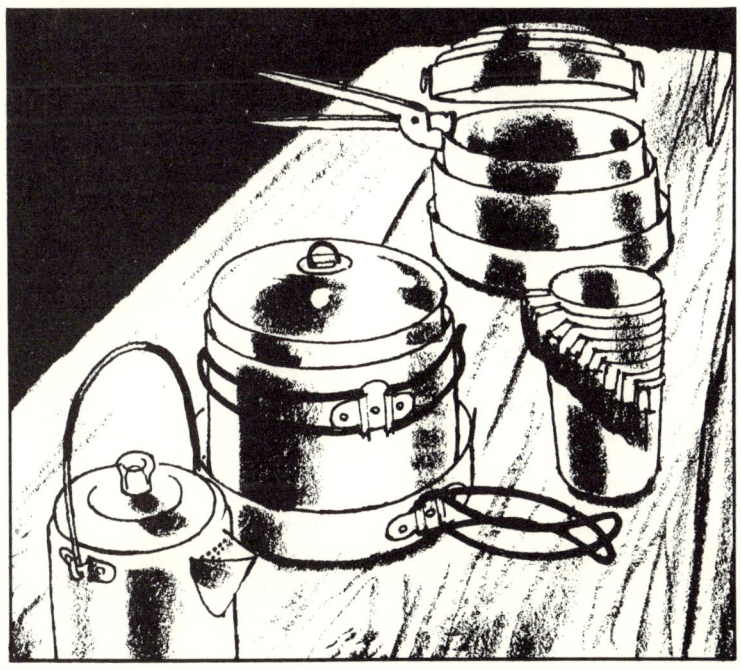

utensils—some are strong enough to use as a seat—and can double as storage on the way home.

Grills are not an essential if you plan to camp in organized areas with stone or concrete fireplaces at each campsite, but you should carry one along, if for no other reason than you will have a backup grill in case the one in the campground is broken, too large, or not even there.

They come in a variety of shapes and sizes. Often you can find one at a second-hand store that came out of a range oven, or you can use one from your barbecue set at home.

Outdoor equipment stores have smaller ones designed for campers that weigh only ounces and have their own carrying case to keep the soot away from other equipment.

The best are those with a fine mesh so you can cook hamburgers and hot dogs on them without losing the meat in the fire.

Griddles are a blessing for pancakes at breakfast and other uses and, like grills, don't take up much space. Some newer models are made of lightweight materials so you won't feel you're carrying a cast-iron stove top with you.

Ovens are available that fit above a burner of your campstove, and most will fold flat when not in use. Most models have a built-in thermometer.

Dutch ovens have long been a favorite for campers, who swear that they are not only versatile and easy to use, but also make food taste better. They are simple to use, and after you've practically buried one under coals for the specified time, the cooked meal that comes from it appears as if by magic.

The true Dutch ovens are sometimes difficult to find, and some outdoor supply stores don't seem to know the difference between them and ordinary kettles. A real Dutch oven is made of cast iron (some new ones are of aluminum, but purists stick with the heavier cast iron). They have three short legs about two inches long (so they can sit slightly above the fire) and a lid that is indented or flanged so hot coals can be placed on it. The lid has a handle, sometimes notched at the exact center of balance, and can be used for frying bacon or sausage and eggs.

New Dutch ovens must be "seasoned" before use. This means you should boil grease, lard, or suet in them, spreading it all over the inside, until it starts smoking. Then remove from the heat (your range oven at home is fine) and wipe it thoroughly—but leave a coating of grease. Never use strong detergent or scouring pads to clean it; you'll have to season it again. After each cleaning, lightly coat it with grease again to keep water from reaching the metal and creating rust.

Dutch ovens can be used for broiling, roasting, stewing, deep-fat frying, and just plain frying on the lid.

Like the crock pots used for home cooking, Dutch ovens are usually used for things like pot roasts and stews, and nearly every river guide in America has a Dutch oven stew he or she has named in honor of a river. One memorable dish, called the Salmon River Stew, has both beef and sausage patties mixed among the vegetables.

For these dishes, the oven and lid are preheated; then grease or suet is dropped in and heated until it begins to smoke and all parts of the oven are swabbed with grease. Then the roast or stew meat is seared on all sides. Add a cup of hot water slowly, salt the meat, and add vegetables.

Some cooks prefer cooking the meat

THE OUTDOOR KITCHEN 125

Dutch ovens are made of cast iron (some aluminum models are available) and have three legs and a recessed lid that holds glowing coals or can be cleaned and used as a skillet.

first, then adding the vegetables about an hour before the meal. Most, however, put the whole dish into the pot and forget about it while pursuing other activities.

There are two ways to heat the oven: over an open fire with coals added to the lid, or buried.

If you want to bury a Dutch oven, four to eight hours, depending on the recipe and the intensity of the fire. After you've completed your day's activities, you can dig out the oven and speed the cooking along with a fresh fire.

Cooking above an open fire can be speeded along by keeping the fire stoked up. For this kind of cooking, use about

A Dutch oven filled with pot roast or stew is surrounded by coals, then buried for a day-long cook.

select a site where there is absolutely no danger of fire spreading or damaging the soil. A sand or gravel bar is best. Scoop out a wide hole, approximately a foot deep, and build a fire in it. When the wood turns into a bed of hot coals, put the oven into the hole, cover the lid with coals, and cover the whole thing with dirt, sand, or gravel. If using sand and/or gravel, cover it with a mound to preserve the heat. Better put some kind of "fence" around it, too, so people won't walk on it.

Buried ovens will cook the meal in the same amount of time you would if cooking on your range at home.

Remember, though, that Dutch ovens are heavy and, if you plan to camp out of sight of your vehicle, you'll probably curse the person who suggested it be brought along.

A twelve- to fourteen-inch Dutch oven will feed up to twenty persons if crammed full of food. A ten-inch size will feed up to ten persons, although campers seem to have heartier appetites than other people, even though some days

CAMPER'S SECRET

If you're thinking of trying a new dish on a camping trip, don't take chances: prepare it before the trip, preferably using the same kind of stove.

they may have done nothing more vigorous than washing the breakfast dishes.

In the recipe section of this chapter, the following dishes are specially designed for cooking in a Dutch oven: Beans, Pot Roast, Corned Beef and Cabbage, and New England Clam Chowder.

Pots and pans: Since most family cooks are rather particular about their cooking utensils, it is smart to invest in a set of nesting cookware and to keep camping equipment separated from the household equipment. Nesting cook sets come in a variety of styles, but most have at least three pots—ranging in size from three quarts to one quart or less. Other larger models are called nesting kettles and have capacities up to six quarts.

Nesting billies differ from other popular sets in that they are not designed to fit on a particular stove, as many cooking sets are. If you plan to car camp only, nesting billies are sufficient.

A typical nesting cook set designed to fit over a particular backpacking stove will include a base and a windscreen on which the three or four nesting pots fit. The whole thing can be placed inside the largest pot, stove included, making a small, compact package.

If you will be cooking over a campfire, almost any old pot and pan will suffice, provided weight is no consideration. You can shop around charity stores for this equipment or use your own second-generation kitchen ware. Usually a maximum of three cooking pots will suffice, and most experienced campers get along very well on two. Use the largest to keep water boiling and the other two for meals. In case of a larger group, all three may be needed. Your menu will dictate this.

After pots and pans you will need the same equipment you use at home—stirring spoon, spatula, good can opener, a sturdy pot gripper or pad, a portable grate, and a sharp knife.

In addition, each member of the party should have a plate, cup, knife, fork, and spoon. Plastic plates are fine, although some prefer using pie plates so they can also be used to cook a cake or cobbler. Cups may be plastic and metal, but some campers prefer to bring along their favorite mug on the grounds that drinks taste "funny" from plastic or metal. A set of tableware that clips together is a wise investment for each member of the party, and the truly fussy might want to label their own set to keep others from using it.

DOING THE DISHES

We have lived with automatic dishwashers so long that they are taken for granted by most Americans, and the thought of actually washing, rinsing, and drying dishes makes many beginning campers—especially children—consider fasting on their camping trip. But doing the dishes is a fact of life in campgrounds and one of those chores that is worse to ponder than to perform.

First, each member of the party should be responsible for his or her own plate, cup or glass, and eating utensils.

CAMPER'S SECRET

Charcoal is always welcome for camp cooking, and you can carry in rinsed-out milk cartons, then burn the cartons to help start the charcoal.

That way, since most camping meals don't require more than one or two pots, the person assigned to the dishwashing duty won't have that many things to wash after each meal.

If you are carrying only one large pot (which is unlikely with the nesting sets being so lightweight and practical), boil water in it, add detergent, scrub each item, and place it on a clean flat surface. When everything is washed, bring another pot of water to boil, then rinse each item thoroughly.

One of the best ways to let things dry is to buy or make a wide-mesh bag for each person that will hold a plate, cup or drinking glass, and all eating utensils. Once washed, equipment can be placed in the mesh bag and hung from a nearby tree or bush to air dry. You can carry a dish towel, which can double as a hot pad, but it isn't really necessary if you can set the utensils out to dry.

To do the washing, you can take along a small plastic bottle filled with dishwashing soap if you want, but those scrubbing pads preloaded with detergent are usually sufficient. Use one for the really dirty jobs, such as scrubbing soot off the bottom and sides of pots placed over the fire, and the other for the insides of pots, pans, and eating utensils. One each of these pads is usually sufficient for a weekend trip (double that for a week's trip).

Since liquid detergent is usually concentrated, a small plastic bottle of it doesn't take much room and comes in handy for soaking pots when the cook chars the stew pot. Buy detergent that is not poisonous; most outdoor stores have biodegradable detergent in small plastic containers or in scrub pads.

Sand from the beach or river bank is a good scouring agent, too, especially when mixed with a bit of detergent. Do not scrub and rinse directly into the stream or lake, however. Although the residue may be biodegradable, you and your neighbors won't like the idea of sharing dishwashing soap, uneaten food, and dirty water. If you aren't at a campground with a wash area and running water, move well away from your camp and your neighbors' campsites into the woods to dump the dishwater. Obviously it should not be dumped on a trail or anywhere someone might step on it.

Most scrapings from pots and pans will burn in the campfire, but be certain it does burn. Don't clean the breakfast egg residue into the fire, then pour water on the fire, leaving behind a mess of blackened wood decorated with fried eggs.

If you camp in an area that does not permit open campfires, carry a sturdy, waterproof sack—preferably nylon—or one lined with a plastic garbage bag and pack out your food garbage with you to the nearest garbage can. Most camping areas that do not permit campfires are

CAMPER'S SECRET

Use sand to scour skillets and pots, or bring pads preloaded with detergents. Dishwater should be dumped well away from sources of water.

THE OUTDOOR KITCHEN 129

heavily used, and it doesn't take long for food scraps dumped in the woods to become garbage. It also attracts wildlife and encourages animals to depend on mankind for food scraps.

MENU PLANNING

Most campers prefer a hearty breakfast a no-bother lunch, and a hearty dinner. This not only gives the cook more time away from the stove or fire; it also helps guarantee that the cook will have lots of help during mealtime; catch the others when they're hungry and they'll be more willing to help instead of wandering off.

In theory, the stove or fire should not

> **CAMPER'S SECRET**
>
> Since most cereal boxes aren't sturdy enough for camping trips, try using well-rinsed half-gallon cardboard milk cartons. These cartons also make excellent water containers, and when you're through with them, they burn easily.

be necessary for lunch, although someone will almost invariably want hot tea, coffee, or chocolate. Some campers have the will power to fill a thermos with hot drinks and consume it only at mealtime, but most—particularly children—do not.

Variety of dishes is important, and you can plan so that each day's menu is different. Eggs with sausage, bacon, or bacon bits can constitute breakfast on

> **CAMPER'S SECRET**
>
> Freeze as much of your food as possible before the trip. This will both keep the food longer and help cut down on the amount of ice you will need on the trip

alternate days, as can omelets. Individual packets of hot breakfast foods, such as oatmeal, can be prepared along with toast made over the fire or stove.

Lunches should always be of the serve-yourself variety: canned meats, paté and "hand foods" such as gorp, high-energy tropical chocolate and fruit bars, pilot bread or crackers, peanut butter and jelly.

Dinner can be the main event: roasts, stroganoff, chili, chipped beef, spaghetti, and so forth. Vegatables—freeze-dried, dehydrated, or canned—should be part of dinner. Desserts can range from cakes baked in camp to pudding, fudge, or even a novelty item such as freeze-dried ice cream. There's no need to limit yourself to boring dishes while camping, as the recipes later in this chapter will attest.

The U.S. Department of Agriculture's guide to daily menu planning can be easily adapted to camping menus. This guide is divided into four basic food groups:

> **CAMPER'S SECRET**
>
> Make your ice chest more efficient during the heat of the day by wrapping it in a blanket, a sleeping bag, or some other insulating material.

THE OUTDOOR KITCHEN

Meat: Two or more servings, for energy and growth. This includes red meats, poultry, fish, eggs, cheese, and legumes.

Dairy foods: Two servings for adults, three to four for children and teens, for growth and body maintenance. This group includes milk, butter, cheese, and other such products.

Vegetables and fruits: One dark green or yellow vegetable and one citrus fruit or tomato.

Breads and cereals: Four or more servings, for roughage. Includes breads, breakfast cereals, macaroni, or noodles and rice.

SAMPLE THREE-DAY NUTRITION PLANNER

Breakfast	First Day	Second Day	Third Day
Protein food			
Cereal			
Fruit or juice			
Beverage			

Lunch	First Day	Second Day	Third Day
Main dish			
Vegetable or fruit			
Bread			
Dessert			
Beverage			

Dinner	First Day	Second Day	Third Day
Main dish			
Vegetable			
Salad			
Bread			
Dessert			
Beverage			

132 FAMILY CAMPING

THREE-DAY MENU

BREAKFAST
scrambled eggs
hash browns
hotcakes
hot drinks
citrus drink

LUNCH
cup of soup
pilot bread or crackers
powdered milk drink
oatmeal cookies

DINNER
beef stroganoff
rice
mixed vegetables
brownies
milk mix
hot drinks

BREAKFAST
granola
cinnamon toast
hot drinks
citrus drink

LUNCH
cheese spread (in tube)
pilot bread
peanut butter
milk drink
coffee or tea

DINNER
macaroni, cheese,
and ham bits
instant potatoes
vegetables
milk mix
hot drinks

BREAKFAST
pancakes with corn
or berries
boiled eggs
toast and jelly
citrus drink

LUNCH
canned meat
bread or crackers
fig bars
fruit drink
coffee or tea

DINNER
vegetable and beef stew
vegetable
applesauce
peach cobbler
milk mix
hot drinks

CAMPER'S SECRET

Here's a general "pinch guide" in case you forget your measuring cups:

2 pinches = 1/8 teaspoon
3 pinches = 1/3 teaspoon
4 pinches = 1 teaspoon
1 fistful = 1/4 cup

Better test these at home in case your fingers are smaller or larger than average. Also test them with your camping spoons and a drinking cup.

CAMPING RECIPES

Obviously your camping menu is limited only by individual tastes and your imagination. Some families pay little attention to menus, other than as a source of nutrition and body fuel, and stock up on the simplest items to prepare, treating eating as a necessity that must be tolerated, like sleep. But most campers like a bit of contemporary civilization combined with their outings, something like the British during the empire days when they dressed for

THE OUTDOOR KITCHEN

dinner even if they were on the muddy banks of the Congo River.

Following is a group of recipes gleaned from a variety of sources to show you things that are possible with only a modest amount of advance preparation and a dependable cooler. These recipes range from the ordinary to the gourmet.

HOT CHOCOLATE MIX

1 lb. instant cocoa mix
6-8 oz. dry cream substitute
8 qts. (or 9 cups) dry milk
1/3-1/2 cup powdered sugar

Mix thoroughly. This fits into a 3-pound coffee can with a plastic lid. To prepare, use 1/3 cup mix to one cup hot water.

HUNGARIAN MEATBALL SOUP

Serves 4

1 lb. lean ground beef (uncooked)
1 cup bread crumbs
1 egg, beaten
salt and freshly ground pepper to taste
1 onion, sliced thin
2 tbsp. butter
2 tsp. paprika
2 cups diced peeled potatoes
garnish: sour cream or plain yogurt and chopped parsley

Combine the beef, bread crumbs, egg, salt, and pepper. Shape into 1-inch balls. In a saucepan, sauté the onion in butter until soft, then add the paprika and 4 cups water. Bring to a rapid boil, lower heat, and add meatballs and potatoes. Cover and simmer 30 minutes. Ladle into soup bowls, top each with a spoonful of sour cream or yogurt, and sprinkle parsley on top.

NUTTY OATMEAL

Serves 2

1 cup oatmeal
1/4 tsp. salt
1/2 cup nonfat dry milk
1 oz. chopped dates
1 oz. chopped nuts (unsalted)
1 oz. brown sugar

Combine oatmeal, salt, and dry milk, and boil in 2 1/2 cups of water. Remove from heat. Stir in dates, nuts, and sugar. Return to low heat and simmer about 3 minutes, stirring occasionally.

NEW ENGLAND CLAM CHOWDER

Serves 6

1 qt. clams with liquor
1/4 lb. salt pork, well rinsed and diced or 2 slices bacon, diced
1 onion, chopped
3 potatoes, peeled and cut into 1/2-inch cubes
3 tbsp. butter
1 pint half-and-half or 1 cup each milk and heavy cream
salt and white pepper to taste

Combine clams and their liquor and 1 quart water. Bring just to a boil in a large saucepan. Drain clams and reserve liquid; chop clams coarsely and set aside. Fry salt pork or bacon just to release fat; add onion and cook until it is transparent. Add the reserved liquid and potatoes and cook covered until potatoes are tender, about 20 minutes. Stir in butter, half-and-half or milk-cream mixture, reserved clams, salt, and pepper. Heat through; do not boil. Serve with pilot crackers.

POT ROAST

While cooking breakfast, braise a roast on all sides, then place it in the Dutch oven with half a cup of warm water, carrots, potatoes, onions, garlic, cloves, salt, and pepper—and whatever else you want. But don't overfill; leave a space between the lid and contents.

When the campfire has died down to a nice bed of coals, dig the coals aside and put the oven in the middle, cover the lid with coals, and cover the whole thing with fresh dirt. Dinner will be served at dusk.

CORNED BEEF AND CABBAGE

Serves 4-6

1 4-lb. corned beef brisket, soaked in cold water to cover for 2 hours to remove excess brine, if necessary. (Can be done before leaving home.)
1 bay leaf
6 peppercorns
1 onion, stuck with 2 cloves
1 carrot, sliced
1 celery rib, sliced
2 sprigs parsley
1 cup apple cider
6 carrots, halved crosswise
6 new potatoes
1 small head cabbage, cut into 4-6 wedges
accompaniments: Dijon-style mustard, horseradish, gherkin pickles

Combine all ingredients, except cabbage, with water to cover and cook in Dutch oven for 8 to 10 hours. Remove corned beef and vegetables to a platter and keep warm. Add cabbage to pot, cover, and cook on high heat for 20 minutes. Place cabbage on platter with meat and vegetables. Serve with accompaniments.

BEANS

Serves 2

1 lb. navy or pea beans
2 chunks of bacon (salt pork)
1 tsp. salt
1/2 cup molasses
pepper to taste

Soak beans overnight, then bring to a boil and cook until the skins burst when spooned or blown. Pour off liquid and save. Drop the salt pork in, add salt, molasses, and pepper to bean water, then pour it over the beans and add the other piece of salt pork on top. Use the hole-in-the-ground Dutch oven method of cooking (see page 126). Cook 4 - 8 hours.

SWISS FONDUE

Serves 4-6

1 clove garlic
2 cups dry white wine
1 1/2 lbs. imported Swiss or Gruyere cheese, grated or cut into small cubes
2 tsp. cornstarch mixed with
1/4 cup Kirsch
pinch of freshly ground pepper
pinch of freshly grated nutmeg
French bread, cut into 1-inch cubes
accompaniments: green salad or fresh fruit for dessert; a dry white wine, such as Riesling

Rub the inside of a cooking pot with the garlic clove, then discard garlic. Heat the wine in the pot placed over fire and add cheese, stirring constantly until smooth. When bubbles begin to appear, add the cornstarch-and-kirsch mixture. Season with pepper and nutmeg. Fondue should be kept slightly bubbling while cooking. Spear bread cubes with a fondue fork, roll in fondue, and eat. Serve with a green salad or fresh fruit for dessert and white wine.

DONUTS

Open a can of ready-to-bake biscuits and cut center holes in each. Fry in shortening until brown, and fry the "holes" too. Drain on paper towels or napkins, then shake in a sack of powdered sugar, white or brown.

CAMPERS' FUDGE

Serves 4

2 cups chocolate chips
3/4 cup sweetened condensed milk
dash of salt
1 tsp. vanilla
1 cup chopped nuts

Melt chocolate over hot water; stir in remaining ingredients. Pour into buttered pie pan, cool, and cut into squares.

CAMPER'S SECRET

Fill quart or half-gallon cartons with water, staple them closed, and freeze them before the trip. They will stay frozen longer than ice cubes and take up less room. The cartons can be used for pitchers later.

SOME-MORES

1 box graham crackers
1 bag marshmallows
chocolate bars

This is a favorite for scouting groups and everyone else who camps. Place a chocolate bar on top of two separate graham crackers. Toast marshmallows until soft, then squash them between the chocolate-lined crackers. You'll undoubtedly want some more.

PEPPER CHEESE

8 oz. cream cheese or pot cheese
2 oz. dry white wine
1/2 pint heavy cream, whipped
1 clove garlic, crushed
1 tsp. finely minced fresh fennel or chervil or summer savory
4 tbsp. peppercorns, crushed

Mix everything except peppercorns in mixing bowl, using fork or electric beater. Chill for several hours, then flatten mixture. Spread the crushed peppercorn on a flat surface and roll cheese in it, pressing down so that the pieces of pepper are pushed into the cheese. Roll the cheese around so that a lot of pepper is coating it. Store in plastic container in cooler until needed.

TOMATO BARBECUE SAUCE
About 2 cups

1 medium onion, chopped fine
1 tbsp. salad oil
1 large clove garlic, minced or pressed
1/2 tsp. dry mustard
1/2 tsp. each salt and chili powder
2 tbsp. brown sugar
3 tbsp. cider vinegar
Worcestershire sauce to taste
3/4 cup each catsup and dry red wine

Cook onion in oil in a 1 1/2- to 2-quart saucepan until soft but not browned; stir in garlic, salt, chili powder, dry mustard, brown sugar, vinegar, Worcestershire sauce, catsup, and wine. Stir until sugar dissolves and mixture begins to boil; simmer for 3-5 minutes, then remove from heat.

Cool before using as marinade for uncooked beef, pork, or poultry. Or brush, warm, over hamburgers and hot dogs.

HOMEMADE GRANOLA
Serves 4

3 tsp. (or more) cooking oil
1/4 cup water
3 cups rolled oats
1/2 cup brown sugar
(or use 1/2 cup honey and leave out water)
1/4 tsp. salt
1/2 cup wheat germ
3 tsp. soy flour

Combine wet ingredients, then dry, and combine mixtures. Roast in shallow baking pan at 210° F. for 25 - 30 minutes, or at 350° for 15 - 20 minutes. After roasting, you may add dried fruit.

For crunchier granola, use only 1 cup rolled oats and add 1 cup unsweetened coconut and 1 cut sesame seeds.

PRECOOKED GROUND BEEF

Many experienced car campers have learned that one of the best shortcuts for a two- or three-day camping trip is to precook two or three pounds of hamburger meat a few days before the trip, then freeze it until the camping trip.

Prepare it the same way you would for chili or shepherd's pie or for a pizza. Fry it in a big pan, adding a bit of water, and cook until totally brown. Drain the water and grease, and place in a plastic container and freeze.

It will keep in the camping cooler for several days, or as long as you keep ice in the cooler, and greatly simplifies your cooking. It can be used in a variety of recipes, including Pizza Buns (next recipe) and Chili Bean Soup (page 137).

PIZZA BUNS
Serves 4

4 tsp. butter
4 English muffins, split
1 medium onion, chopped fine
1 15-oz. can tomato sauce with bits
4-oz. can mushroom stems and pieces, drained
1 tsp. Italian herb seasoning
1/2 tsp. garlic salt
1 lb. precooked ground beef (see preceding recipe)
4 oz. Jack cheese, shredded

Melt 2 tsp. butter in a 10-inch fry pan, and toast muffin halves. Keep warm. Melt remaining 2 tsp. butter in pan and sauté onion. Then stir in tomato sauce, mushrooms, seasonings, and ground beef. Cover and cook, stirring occasionally, until heated. Spoon over muffin halves, then sprinkle with cheese.

COLESLAW

Serves 6-8

3 cups shredded cabbage (1/2 head)
1 cup shredded carrot
1/2 small red onion, peeled, thin-sliced, and separated into rings
1/2 cup sour cream
1 tbsp. sugar
2 tbsp. tarragon vinegar
1/2 tsp. salt

Combine cabbage, carrot, and onion with sour cream, sugar, vinegar, and salt in large bowl. Toss the slaw to coat thoroughly, and chill until eaten.

BATTER FOR HOT BUTTERED RUM

1 lb. butter
1 lb. brown sugar
1 qt. vanilla ice cream

Soften butter and ice cream and mix thoroughly. Makes 4 pounds of batter that can be stored frozen and used as needed.

Servings: To 2 tablespoons (or more) of batter, add rum to taste and hot water. Children love the batter (without the rum, of course).

CHILI BEAN SOUP

Serves 4

3 tsp. butter
onions to taste
30-oz. can chili beans
28-oz. can tomatoes
1 lb. precooked ground beef (see page 136)
2 tsp. dehydrated sweet pepper flakes
1 tsp. garlic salt
2 - 3 tsp. chili powder

Fry onion in butter. Add rest of ingredients, cover, and simmer for 15 minutes.

CORNED BEEF AND NOODLES

Serves 5-6

2 12-oz. cans corned beef
12 oz. egg noodles
3/4 cup catsup
2 tsp. Worcestershire sauce
dash of Tabasco sauce
2 tsp. vinegar
1 cup water
1 tbsp. minced onion (or instant onion)

Boil egg noodles according to package directions. Combine sauce items separately and simmer about 5 minutes. Then break up corned beef and add to sauce, and simmer about 15 minutes. Drain noodles and mix with sauce and beef.

BEEF FONDUE

Serves 4-6

1 1/2 - 2 lbs. boneless top round
unseasoned powdered meat tenderizer
2 cups salad oil
salt and pepper
Spicy Garlic Mayonnaise
(see next recipe)

Trim fat from meat and cut into small cubes. Sprinkle with meat tenderizer and let stand for about 30 minutes, following tenderizer directions. Heat oil in pot to about 360° F. If you're using a campstove, lower burner and keep oil hot; if you're carrying a canned heat burner, use that. Cook each piece of meat to taste; 20 - 30 seconds will brown the outside, leaving the meat rare inside. Transfer meat to serving platter, sprinkle with salt and pepper, and dip into Spicy Garlic Mayonnaise for serving.

SPICY GARLIC MAYONNAISE

About 1 1/2 cups

1 egg
2 tsp. Dijon-style mustard
1 tsp. paprika
1 clove garlic, minced or pressed
1/2 tsp. salt
2 tbsp. white wine vinegar
1/4 cup olive oil
3/4 cup salad oil
1 tbsp. each chili sauce and drained capers
1 tsp. Worcestershire sauce
2 tbsp. snipped fresh chives or thin-sliced green onions
2 tbsp. chopped sour pickle

Before leaving home, combine in blender egg, mustard, paprika, garlic, salt, vinegar, and olive oil. Cover and blend at low speed, then immediately pour in salad oil in steady stream, and whirl until thick and smooth. Add remaining ingredients. Cover and keep chilled, then store in cooler for camping trip.

TOMATO RAREBIT

Serves 4

1 can condensed tomato soup
1/3 can milk
1/2 lb. (approx.) American cheese (sliced or diced)
1 egg

Heat tomato soup, milk, and cheese until cheese is melted. Beat egg, stir slowly into the cooking mixture, and cook for one minute. Serve over toast or crackers.

MARINADES

A variety of marinades can be made at home or from mixes bought in grocery stores, and experimenting at the backyard barbecue will tell you which is the most popular with your family. Marinating is one of the best ways to make inexpensive meat palatable.

One of the simplest marinades is made of two parts cooking oil and one part vinegar. Other recipes with more flavor to them follow:

MUSTARD AND HERB MARINADE

Less than 1 cup

1/3 cup salad oil
1/2 cup dry white wine
1 tbsp. each red wine vinegar and lemon juice
1 large clove garlic, minced or pressed
1 1/2 tbsp. Dijon-style mustard
1/4 tsp. each salt and sugar
1/8 tsp. each thyme, oregano, summer savory, and tarragon
dash of white pepper to taste

Combine all ingredients in blender until smooth. Use for lamb or chicken.

TERIYAKI MARINADE

About 3/4 cup

1/2 cup soy sauce
3 tbsp. sugar
2 tsp. grated fresh ginger or 1/2 tbsp. ground ginger
1 clove garlic, minced or pressed
2 tbsp. dry sherry

Mix or shake ingredients well. Use for beefsteaks or chicken.

SHISH KEBAB

Serves 6

1 lb. stew meat
1 16-oz. can pineapple chunks
1 16-oz. can potatoes
1 onion
1 16-oz. can cherry tomatoes
1 green pepper

Cut meat into small chunks and place in mixture of two parts cooking oil and one part vinegar. Let it marinate at least 24 hours, either before leaving home or in a plastic jar with a screw lid, left in the cooler at camp until needed.

Place meat on skewer, green stick, or spit, alternating with pineapple and vegetables. Cooking time is approximately 15 minutes over normal fire or coals. You can brush kebabs with barbecue sauce if desired.

SNACK ITEMS

Gorp: Acronym for Good Old Raisins and Peanuts. Also added are items such as M&M candies because they keep longer. Can be purchased as a mix or made at home, substituting other nuts, your choice of raisins, and other items.

Fruit: Usually freeze-dried or dehydrated for ease of handling and long life.

Trail cookies: These are high in nutritional value and proportionately expensive. Oatmeal cookies with fruits added are a good homemade substitute.

Party mix: This can either be purchased ready made or made at home. Combination of pretzels, Cheerios, nuts, and Wheat or Rice Chex, baked in the oven, and packaged in individual portions or in a large container.

Pilot bread: Large round crackers that don't crumble as easily as saltines.

EXTRAS

Bacon bar: A compressed bar of precooked bacon, excellent for adding to scrambled eggs or munching for snacks. Has long life.

Meat bars: Made of a variety of meats in same way as bacon bars, and won't spoil easily.

Pemmican: This old Indian recipe of meats and berries has been updated into individually wrapped bars that can be eaten like candy, or in cans and plastic containers to be used as a spread.

WILD FOODS

Every region of the world has its own wild edibles there for the taking—salad greens, nuts, fruits, roots, mushrooms, berries. Since the species vary so much from region to region, we won't attempt a complete listing here, but accurate and detailed guides to these wild foods are available in most outdoor equipment stores and regular bookstores. Look in the regional section of bookstores.

One word of warning: be absolutely certain you know your mushrooms before supplementing your food supply with those found in the forest. A few species are poisonous and some are deadly. Be

---

CAMPER'S SECRET

You can prepackage your camp syrup by combining one-half cup each of white and brown sugar with a dash of cinnamon. In camp, add one-third cup of water and simmer until dissolved.

certain you can identify them beyond a doubt.

Here are a few of the more common plants found in most parts of North America that can liven up your camping meals:

Cattail: One of the most abundant of the aquatic plants, it is also one of the most popular for wild-food gatherers. Just below the leaves is the tender shoot that tastes something like a cucumber and may be eaten raw or cooked. The root is also quite tasty and can be cooked like a potato, fried, baked, boiled, or mashed.

Chickweed: This bane of lawn owners can be partially controlled by eating it. Its leaves and tender stems can be used as a salad or a cooked green.

Dandelion: Another lawn pest, this one has been popular not only for eating but as an ingredient for wine. The leaves may be eaten as a salad, but the younger leaves should be picked to avoid the bitter taste that comes from older leaves. The roots can be chopped and roasted until hard and brittle, then used as a coffee substitute.

Dock: This common weed is known also as sheep sorrel, Indian tobacco, and curly dock. The leaves can be cut and boiled for about five minutes and taste a bit like asparagus.

Lamb's quarters: Usually found in drier climates, this plant is best as a cooked green. Some people call it pigweed, but it should not be confused with the true pigweed, which doesn't taste good at all.

Stinging nettle: Few things are more uncomfortable than a patch of your skin that has brushed a stinging nettle. In spite of this, by carefully pick-

Dandelion

CAMPER'S SECRET

Jar-lid rings can be used to hold the shape of poached eggs. Place the rings in two inches of water in the Dutch oven, then break eggs into them.

ing some of the leaves (preferably with gloves) and boiling the leaves a minute or two to remove the stinging oils, you can sample one of the tastiest of the wild plants. Boiled, it tastes much like spinach.

Bear grass: Usually found in higher elevations, even alpine, bear grass does grow in lower elevations and is easily identified by its tall spike of white flowers. The root is excellent eating and is pulled from the ground and peeled. It can be eaten raw or added to a stew like any other vegetable.

Arrowhead: Sometimes called a duck potato or tule potato or by its original Indian name of Wapato, this aquatic plant has underwater tubers that are excellent eating. It tastes something like potatoes and was used as a medium of exchange by many Indians.

CAMPER'S SECRET

Before the trip, fill plastic jugs with fruit drink or lemonade and freeze them. On the trip, these will help keep everything else in the cooler cold and will give you cool, refreshing drinks for two or three days.

CHAPTER 9

CAMP ACTIVITIES

Camping should be an active form of recreation, not just changing the places you sit on a weekend or vacation. Part of your preparation for the trip should include planning activities at the campsite that you normally wouldn't do at home. Camping is hardly the place to catch up on homework or correspondence, but it is a perfect opportunity to learn more about the outdoors.

The relatively simple act of going somewhere, pitching a tent, cooking meals, sleeping, breaking camp, and returning home is not sufficient cause for many families to consider camping as a goal in itself. However, there are many activities people engage in that could be enhanced by camping. These include hunting, fishing, bird watching, rockhounding, bike touring, canoeing and kayaking, cross-country skiing, geological field trips, photography trips, horseback treks, and summer festivals.

In each of these activities, and many others, basic camping skills and good lightweight equipment are essential to getting the most out of the trips. The more expert you and your family become at camping, the more adjunct activities you will find.

An example is the variation on the license-plate game children love to play while traveling. Instead of tallying up the number of licenses from other states, take along a guidebook that tells you how to identify species of plants, trees, flowers, and birds of the region you're camping in. Keeping a tally of species is a good way for the whole family to learn more about nature. One of the best

series of books on identification, with regional divisions, is published by the Audubon Society.

If you live in the city, the combination of too much light at night and air pollution shuts off your view of stars. But out in the woods they shine vividly from the deep sky overhead. Most bookstores and some outdoor supply stores sell star identification books or charts that can keep the children busy most nights attempting to identify the important stars and constellations. A side benefit of this is that they also learn the basics of celestial navigation.

CAMP ACTIVITIES 145

Another practical activity is learning to use a compass and topographical maps. The basics of navigation can be learned by any child old enough to read and do simple mathematics. Learning to read the colored topographical maps is intriguing to children, for it appeals to their innate imagination. Often it is easier for them to mentally transfer those squiggly lines on maps to the surrounding terrain than it is for adults. As with learning to use the compass and the basics of celestial navigation, knowing how to read the topographical maps may one day be a life saving skill.

Navigation is fun and lends itself well to campground games. The child who thinks math is the pits will often be the one who pores over the maps and compass by the hour. A topographical map is as real as a photograph, and a compass is in some ways the ultimate toy. The stars, too, are real and fun to discover. Once a child learns to identify the major constellations, it helps make those nights a little less overwhelming.

Obviously not all campground recreation must be in the educational field; schoolchildren will quickly rebel if camping begins to sound suspiciously like a school field trip.

So don't hesitate to bring along cards, popular games, small chess or backgammon boards (some are tiny and magnetized), or whatever the family likes to play at home. These help ease long afternoons when the rain is falling or evenings around the campfire or in the light of a lantern.

TOPOGRAPHICAL MAPS

The accompanying illustration shows the basic components of these maps that are the standard for outdoorsmen all over North America. The only way to become accustomed to them is to use them in the outdoors. Nothing replaces practical applications, and while you can pore over them for hours at home to select potential campgrounds or hiking trails, still you must consult them in the actual terrain to learn how to read them at a glance.

These "topo" maps are color-coded. Features such as buildings, roads, railroads, mines, windmills, churches, and schools are all printed in black. Water

CAMPER'S SECRET

When you purchase maps for camping trips, whether in the wilderness or along major highways at established campgrounds, consider buying two of each map: one for use on the trip and another for posting on a wall at home with the route marked in colored ink. As you return to the same area and visit different places, use another color ink. Such maps are nice mementos of the trips, and they also help teach map reading to children.

features are in blue. Vegetation is in green, and all the elevation indicators such as contour lines, altitude markings, and benchmarks are in brown.

Nearly all maps have detailed instructions printed on them along with a key to the symbols.

The standard map for outdoors enthusiasts is on the scale of 1:24,000, which means one unit on the map (this may be a kilometer or a mile or whatever has been chosen and noted) equals 24,000 of the same units in actuality. For example, a mountain that is shown on the map no larger than a pencil eraser will actually be 24,000 times as large.

Each topo map also shows the declination correction factor, which is vital in using your compass. This will be explained in the next section.

Topo maps are available at nearly every major outdoor supply store, at map stores, and by mail order from the U.S. Geological Survey. (See page 213 for addresses.) If you know the specific map or maps you need by their quadrangle names, fine. But it is best to write and ask for an index map, which shows the whole United States, then breaks it

CAMP ACTIVITIES 147

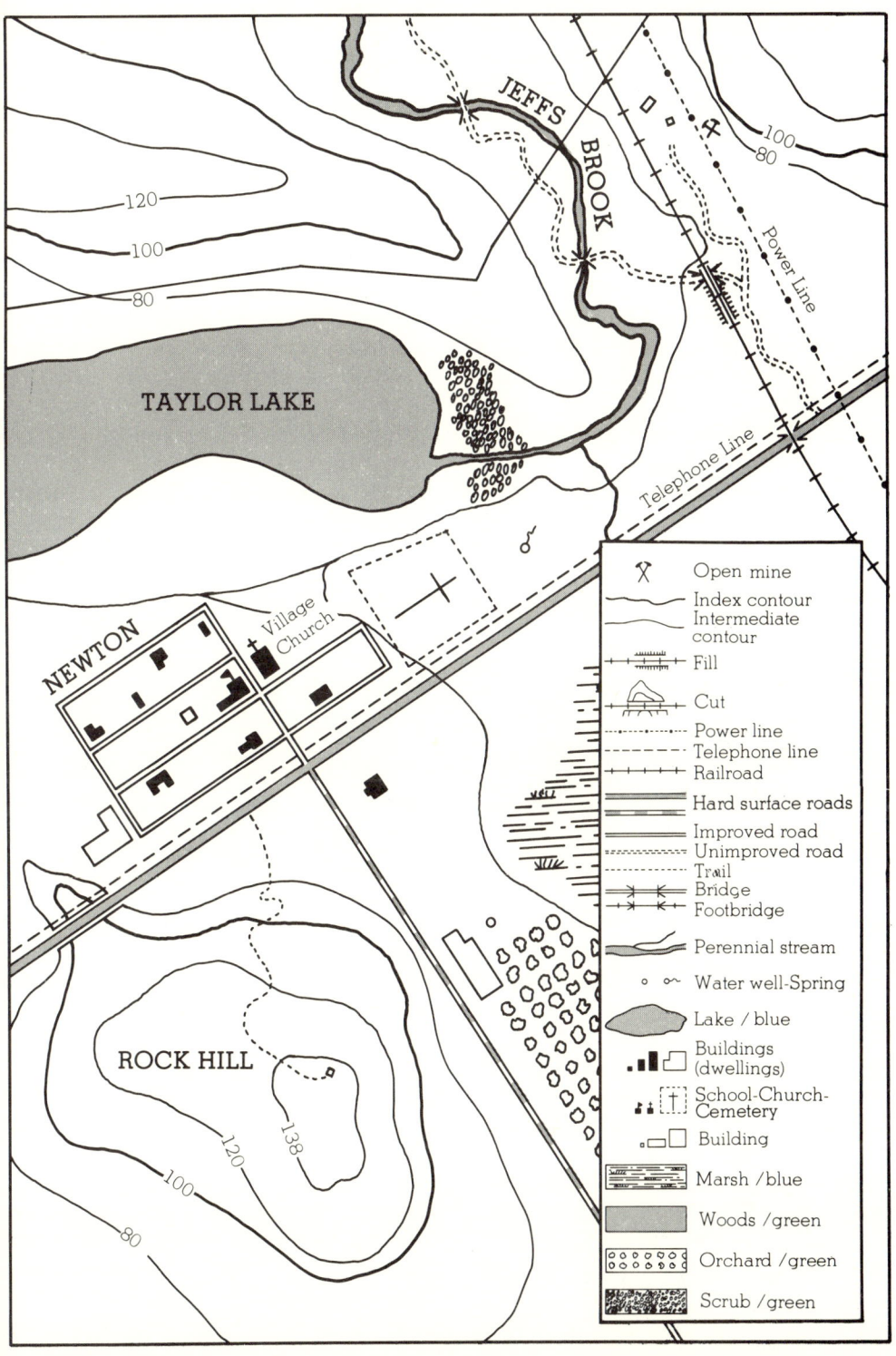

down into increasingly small portions, or quadrangles, so you can order exactly what you need.

Almost identical maps are available for all of Canada through the Geological Survey of Canada.

It is essential to keep topo maps up to date. National Forest roads are continually being built, obliterated, or extended. Buildings shown on maps burn or vanish. New ones are built. New landmarks, such as microwave towers and power transmission lines, are built. So it is best to order your topo maps directly from the USGS if possible, since outdoor stores might have an older edition that hasn't sold out.

THE COMPASS

The compass is a needle magnetized on one end and balanced on a pin so it can swing freely. The magnetized end will always point toward the magnetic north pole unless it is deflected by local ore deposits (a rare occurrence) or metallic objects you are carrying.

The magnetized needle is mounted above a dial that shows all the major directions and often is marked off in the 360 degrees of a circle as well. Covering most compasses is a plastic plate that can be turned, and it is printed with an arrow marked north.

A compass does not point toward

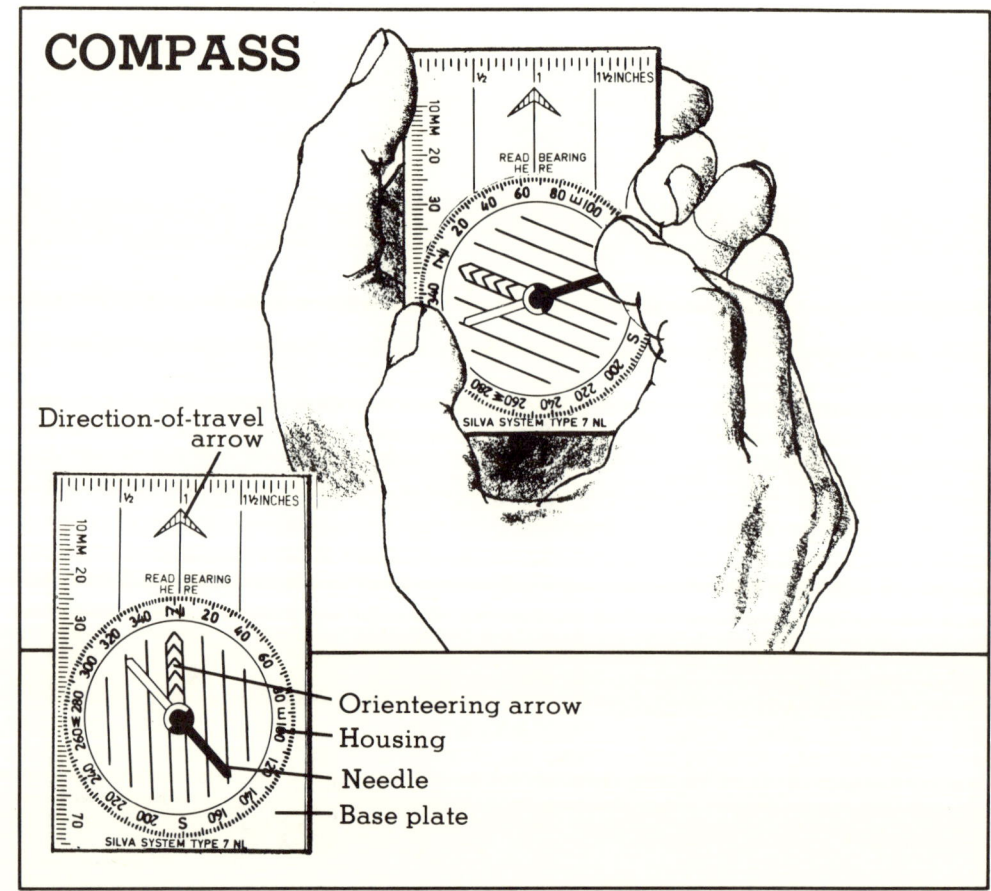

the North Pole. It points toward the magnetic pole, which is approximately 1,000 miles from the North Pole in northern Canada. This magnetic field tends to wander slightly, but not enough to make a lot of difference in your direction finding. At the most, it wanders only a degree.

All maps are based on the North and South poles. They are the only constants from which cartographers can operate with total accuracy. This means that your compass and the maps you carry do not match: whereas the longitude lines on the map point north, your compass will point a few degrees in another direction, toward the magnetic pole.

The angle of difference between the North Pole and the magnetic north pole on your compass is called declination. This varies in different parts of North America, and all U.S. and Canadian Geological Survey maps have the angle of declination printed on them.

This is where the plastic covers of the best compasses are needed. When you plan to use your compass on a trip, you must first consult the map of the area you will be traveling and find the angle or degrees of declination on it. Then you hold the compass with the arrow pointing to north on the dial, count the degrees of declination in the direction away from true north, and turn the upper dial to that point. Thus, the printed arrow will always point to the North Pole while the magnetized needle points to magnetic north.

This is the basis of all navigation by map and compass. The rest is applied mathematics and is relatively simple, provided you learn navigation in easy stages, starting in your own backyard or a local park. If you think you may need to use a combination of map and compass on a trip, it is vital that you practice with them before the trip, and that practice should be thorough and often enough so that you are as comfortable with the compass as you are with street maps.

In most family camping situations, you will not need the compass because you will probably want to stay near the major trails and campgrounds. But these family camping trips are excellent opportunities to teach everyone in the family the basics of wilderness navigation.

As your family becomes more adept at using the compass and maps, you may want to invest in more complex compasses. There are several different kinds available, each with its own uses. Of the five basic types, the **fixed-dial** compass has already been described. Then there's the **lensatic** compass equipped with a dial that pivots beneath the magnetized needle and has an azimuth scale that also is adjustable.

The **cruiser,** designed for forestry engineers and timber cruisers, has a dial with the degrees printed counterclockwise. It is of little use for recreationists.

The **sighting** compass is a sophisticated and highly accurate hand-held compass with a sighting lens in the case that magnifies the dial to within one-half-degree readings. They are useful for canoeists who want to maintain a totally accurate course and for other similar uses.

The **orienteering** compass is a sophisticated design that can be laid on maps for plotting routes. It has a transparent plastic plate for a base that is marked on one side with the map scale in millimeters; another side has the scale in inches. The orienteering compass is a good second choice for the family who wants to combine U.S. Geological Survey map reading and compass use in the same learning process and should

SUMMER CONSTELLATIONS
Regulus
LEO LION
URSA MAJOR
BIG DIPPER
Mizar
DRACO
DRAGON
LYRA LYRE
Vega
CYGNUS
SWAN
URSA
MINOR
LITTLE DIPPER
Polaris
POLE STAR
GEMINI
TWINS
Pollux
Castor
CEPHEUS
AURIGA
CHARIOTEER
CAPELLA
CASSIOPEIA
LOOKING NORTH
HORIZON

HERCULES
BOOTES
Arcturus
CORONA
NORTHERN
CROWN
Altair
AQUILA
EAGLE
OPHIUCHUS
SERPENT BEARER
VIRGO
VIRGIN
LIBRA
SCALES
Spica
Antares
CORVUS
SAGITTARIUS
ARCHER
SCORPIO
SCORPION
CENTAURUS
LOOKING SOUTH
HORIZON

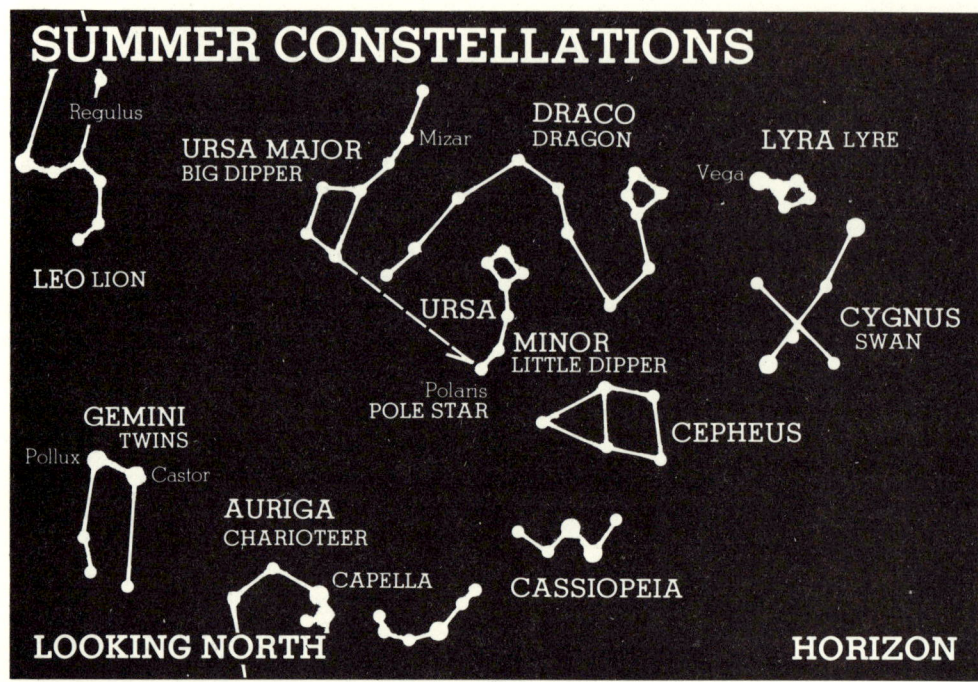

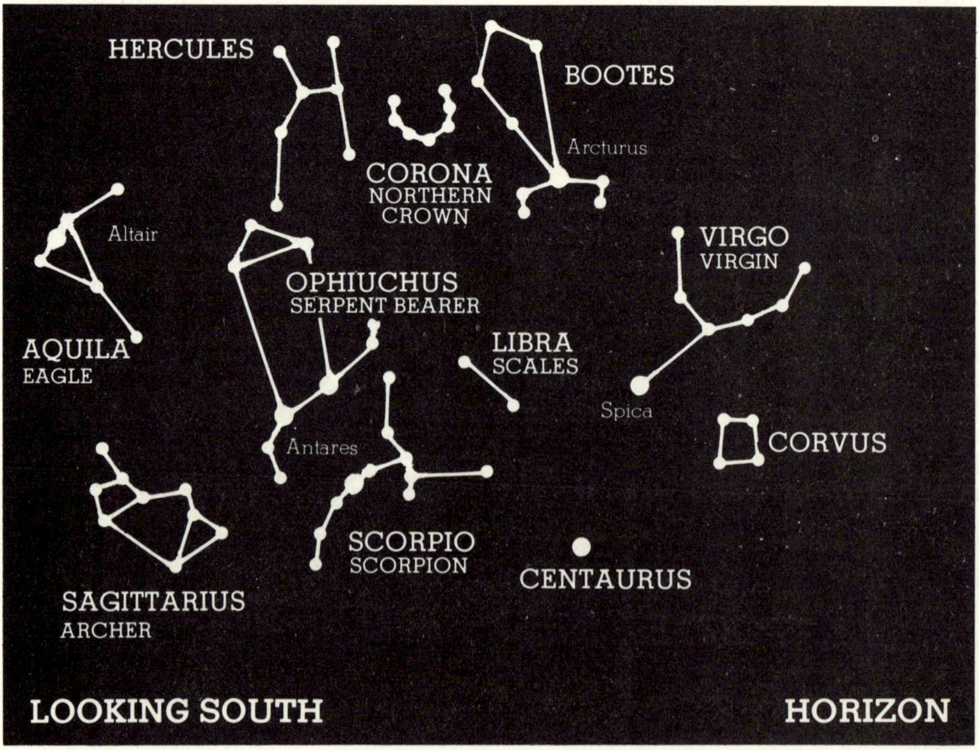

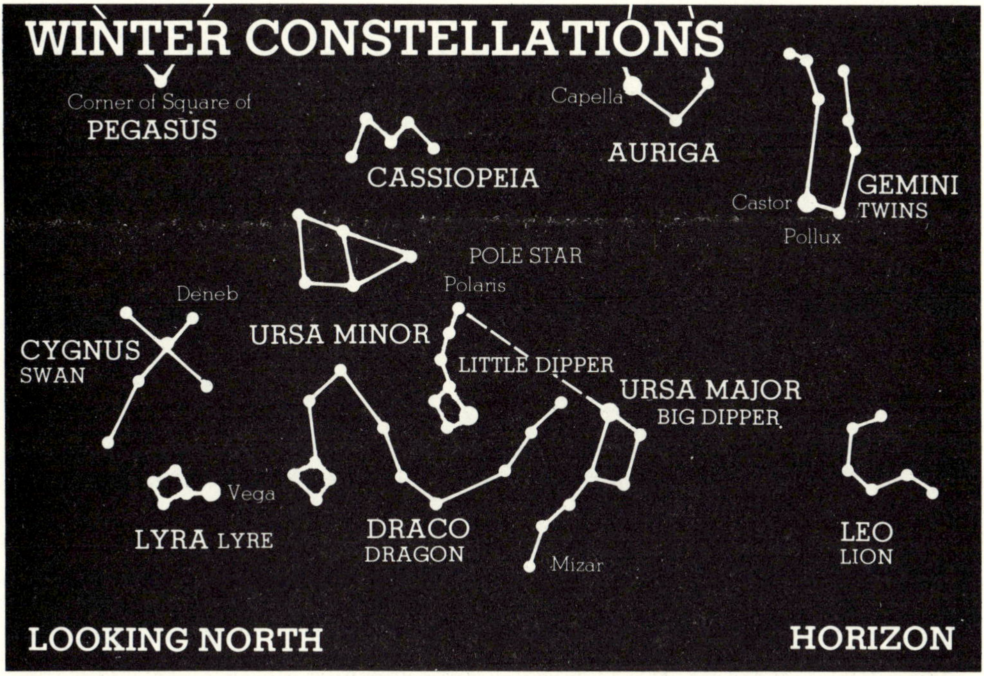

CAMP ACTIVITIES 151

WINTER CONSTELLATIONS

LOOKING NORTH — HORIZON

- Corner of Square of PEGASUS
- CASSIOPEIA
- Capella — AURIGA
- Castor, Pollux — GEMINI TWINS
- Deneb — CYGNUS SWAN
- URSA MINOR — POLE STAR Polaris — LITTLE DIPPER
- URSA MAJOR BIG DIPPER
- Vega — LYRA LYRE
- DRACO DRAGON — Mizar
- LEO LION

LOOKING SOUTH — HORIZON

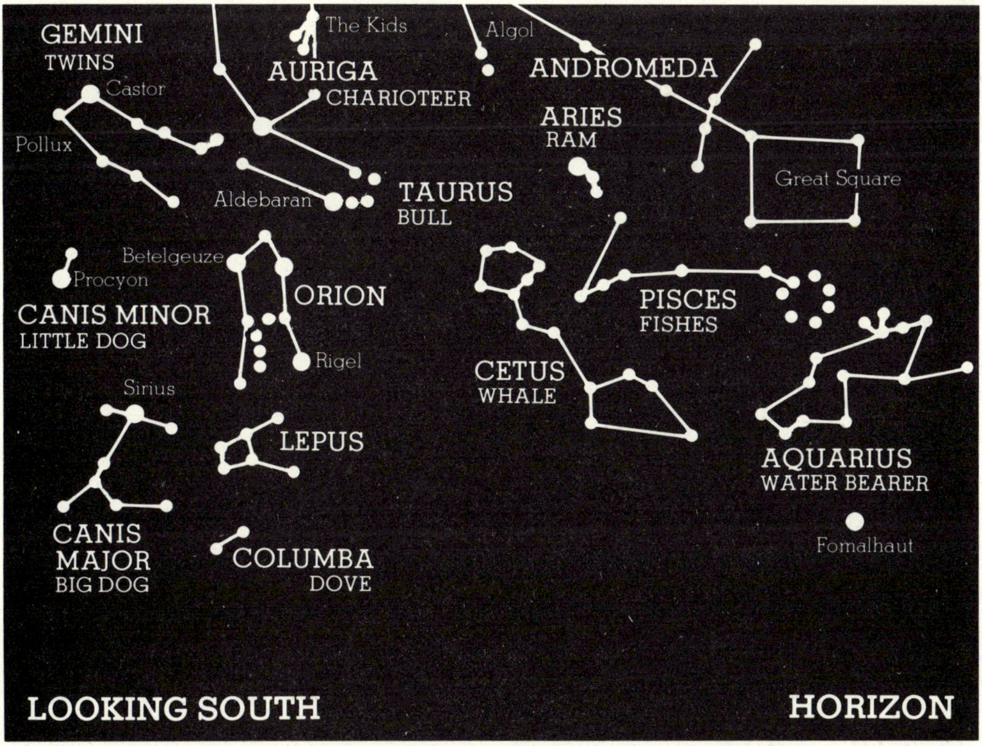

- GEMINI TWINS — Castor, Pollux
- The Kids — AURIGA CHARIOTEER
- Algol — ANDROMEDA
- ARIES RAM
- Great Square
- Aldebaran — TAURUS BULL
- Betelgeuze, Procyon — CANIS MINOR LITTLE DOG
- ORION — Rigel
- PISCES FISHES
- CETUS WHALE
- Sirius — CANIS MAJOR BIG DOG
- LEPUS
- COLUMBA DOVE
- AQUARIUS WATER BEARER — Fomalhaut

be considered as the second step up in compass purchases.

ORIENTEERING

This sport has become one of the most popular forms of recreation in many parts of North America, and especially in Europe. Essentially, it involves someone laying out a course with compass and topographical map for others to follow. The game is one of precision because you must follow the compass and map exactly, and it is one of speed because the first participant to complete the course wins.

This can be adapted for a campground game with checkpoints near enough to the camp so that nobody wanders off into the wilderness. It is a technical version of a scavenger hunt and one of the best ways to become intimately familiar with maps and compass.

A modest version of orienteering can be developed for the whole family as a learning tool. Simply lay out a course along the road to camp and have someone in the group read off the landmarks, follow the compass readings, and alert the driver which landmarks to watch out for next.

Another good way to become well acquainted with both map and compass is to find a camping spot in an open valley surrounded by peaks or hills with some other landmarks on the topo map. One member of the group can go out alone and establish checkpoints along a route, such as hiding a coin (or any recognizable item) at the base of a tree. Obviously the course must be simple for the beginners, and it should specify how many feet or kilometers each leg of the course is. This also teaches campers to estimate how many paces they must take for specified distances, such as one hundred yards or one hundred kilometers.

Another exercise can be conducted in wide-open country the first time, then in timbered and rugged terrain after a bit of practice. This is how to find a spot, such as a campsite, by making progressively smaller boxes on an imaginary map.

Assume you know roughly which direction you've walked from camp, but you know you can't reach it by walking a straight line. So you estimate how far you've walked. Then, from your present position, begin walking at a right angle from the direction you think camp lies. Walk at least half that distance at the right angle, then make a right-angled turn and walk that distance again, then another right angle, which will put you in a straight line from where you started.

Now, walk toward the starting point, but stop three-quarters of the way there (if you've walked one mile, stop at three-fourths) and take your right-angle turn inward again. If you don't see the camp on this leg, make another right-angle turn three-fourths of a mile along, then repeat it until you've made another three-fourths box.

Now cut your distance of the next leg another fraction.

Be sure that while performing this exercise you follow the compass route as closely as possible and allow for detours around rock outcroppings and other obstacles to your direct route.

Unless you are totally lost, or you didn't bother to mentally note landmarks around your campsite, you should pass it before the series of boxes you're walking come to the center. If you don't find your campsite in this fashion, begin another series of progressively smaller

CAMP ACTIVITIES 153

boxes farther along the backward route.

Obviously, an experienced outdoors person will mentally establish a series of landmarks along the route away from camp. However, hikers sometimes get caught in a dense fog that hides everything within a few feet.

It is virtually impossible for anyone to walk in ever-smaller circles, but with a compass to establish a route and reasonably accurate estimates of distances walked, this will bring you back to camp.

WATCH COMPASS

A good trick for everyone in the family to learn (although your chances of having to use it should be slight; always carry a compass) is how to find true north by using your pocket or wrist watch.

First, if you are on daylight saving time, move the watch back an hour to sun time.

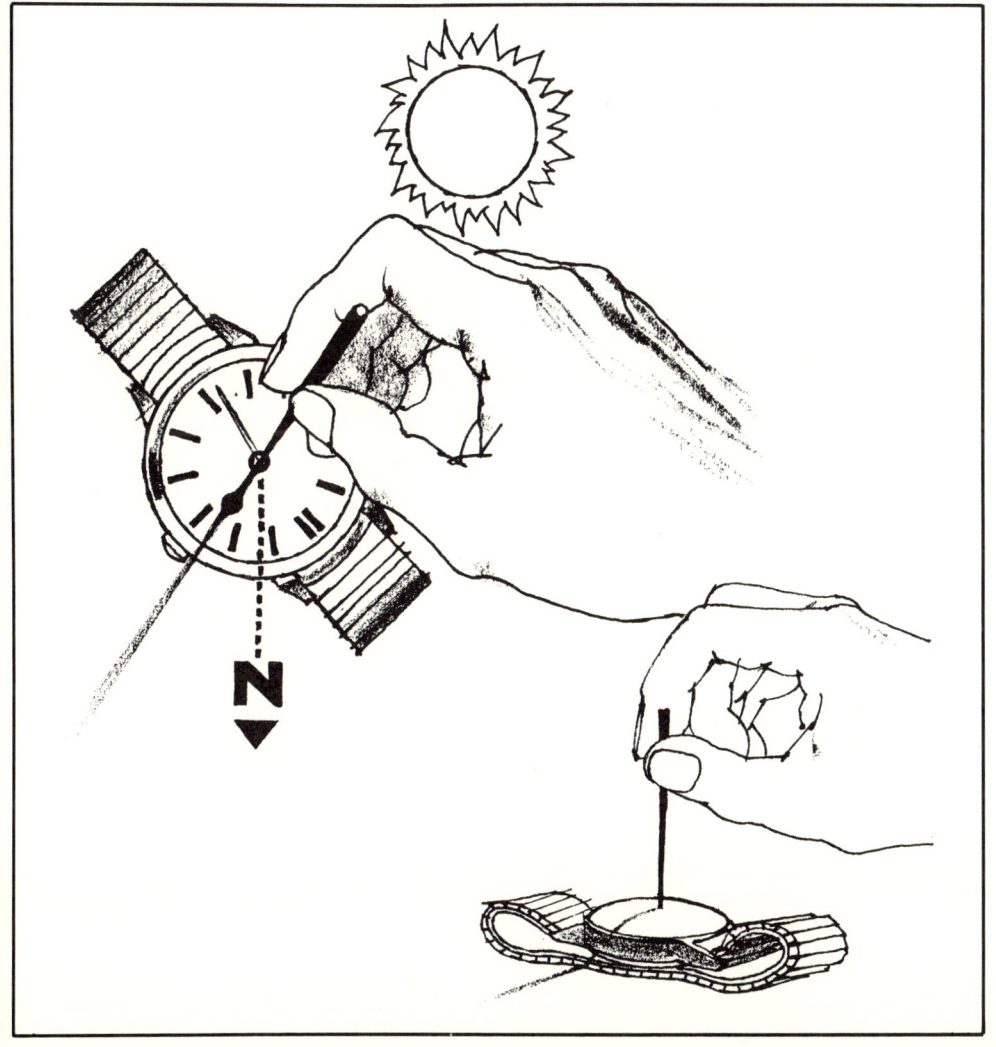

Then get a piece of straight stick—a wooden matchstick will do fine. Place the watch on a level surface, such as a stump of rock, with the sun hitting it. Hold the stick so that it is in the center of the watch, and turn the watch until the shadow of the stick is along the hour hand.

If you are performing this exercise between 6:00 A.M. and 6:00 P.M., **south** not north will be directly between the hour hand and 12:00 on the watch in the angle formed by the two hands.

If you're doing this between 6:00 P.M. and 6:00 A.M., **north** will be between the hour hand and 12:00.

The formula to remember is **N** for North and Night; **S** for South and Sun.

This is accurate to within eight degrees, assuming your watch is set properly.

Another method of finding directions with a watch is to put a straight stick in the ground and mark the end of its shadow at 10 A.M., then again at 2 P.M. (You can use other times, but each must be the same number of hours away from noon.) Measure the distance between the two points, divide by half, and that point will be true north.

CAMPER'S SECRET

Here's how to measure the distance between you and that last bolt of lightning. When the lightning strikes, count off the number of seconds between the strike and the arrival of the thunder, and figure four and a half seconds per mile. If you don't have a watch, count slowly, "one thousand, two thousand," etc.

NATURE SCAVENGER HUNT

A scavenger hunt is always popular with young children and can be used as a method of getting them acquainted with a camping area. Adults should go out first and scout the area not only for items for the children to find, but also for potential dangers. After scouting the treasures—draw up a list that looks something like this:

a pine cone off the ground
a bird's feather
a wild flower
a dead leaf
a living leaf
one piece of litter
one piece of moss off a tree
one piece of moss off the ground

Make sure first, though, that no poison plants such as ivy or oak are in the area and do not let children stray beyond the campground perimeters. If there are several children in the group, divide them into teams.

The winner gets a prize, an extra dessert treat or whatever you choose.

KNOTS

It isn't of great importance in the field to know the language of knots, but when you look through a selection of knots, you will find them divided into knots, bends, and hitches.

A knot is used to tie a bundle of something and it makes a loop or a noose or knot in the rope.

CAMP ACTIVITIES 155

A bend ties two ropes together.
A hitch ties a rope to something—a tree, a ring, an oar, or whatever.

Square knot: The most common, after the slipknot you use to tie your shoes, is the square knot. It won't slip if tied correctly, and it can be untied with considerable ease.

Bowline: This is used for strong loops that won't jam or slip. To tie it, form the loop, then run the free end through, around the standing part, and back through the loop.

Sheet bend: This is used to join two ropes of different sizes. The free ends must be on the same side of the knot or it will slip. This knot will jam on heavy loads, however, and the Carrick bend should be used for those.

Carrick bend: Virtually jamproof, this can be used to join two ropes of different sizes when a strong pull, as with a winch, is needed. To tie, make the loop with one rope, then interlace the other rope as shown. Since there will be considerable slippage until the knot is tight, allow plenty of length on each end.

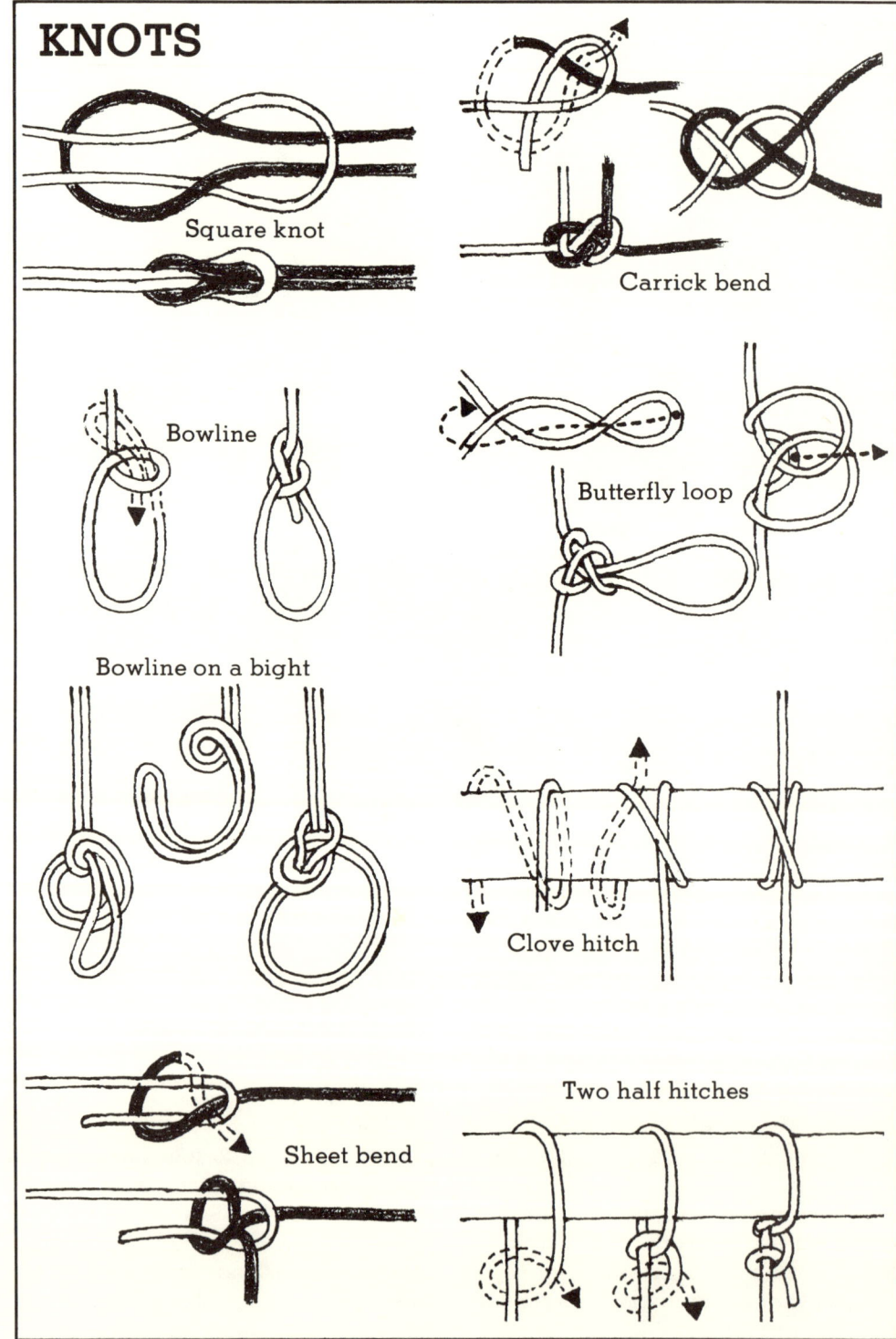

Butterfly noose: This is a loop in the middle of a rope and will not jam. To tie it, twist a loop, then fold up the lower part of the loop and push it through the center opening as shown.

Clove hitch: This is a quick and secure method of attaching a line to a tree or post. It will, however, slip unless pressure is kept on the end. To avoid this, take a half hitch with the loose end around the taut part of the rope.

Half hitch: Half hitches and double half hitches are popular among sailors and cowboys for making quick, temporary knots.

SURVIVAL SKILLS

A number of wilderness survival skills can be taught as a normal part of the camping experience, and while some of this lore is absorbed almost by a process of osmosis, there are many things that can be a part of the camping trip.

Some old legends never die, such as the alleged ability to find north by seeing which side of the trees have moss growing on them. This may be true in some areas of North America, but not all. Some shaded forests have almost equal amounts of moss on all sides. In general, the south side of hills and mountains have less plant growth than the shadier and damper north side. In some areas of extreme glaciation during the ice ages, the hills have been scoured in a general north-south direction.

Animals are no more eager to waste energy than people are, so well-worn game trails usually follow the path of least resistance.

When you are traveling any distance in the outdoors, teach your children—and learn this yourself—to stop often and look behind you so you will recognize the route on your return. Things always look different when approached from the opposite direction. It is especially important to stop and study the lay of the land when you come to a fork in the trail. Memorize the intersection by looking backward at it.

Some outdoors enthusiasts insist they have body chemistry that makes them know instinctively where north is. This may be true in some extremely rare cases, but it is more likely that they have traveled in the outdoors so long that they subconsciously note wilderness signs and know their direction. Without a compass, nearly everyone traveling on strange terrain will walk in a wide circle.

If you are camping where wildlife is abundant, or if you're camping in the off season when snow is on the ground, a book on animal tracks is a good investment. You can also identify animals by their droppings. Soon you will learn to tell approximately how long ago the animal was there by a variety of signs, such as water seeping into the tracks, or crushed grass still returning to its original position.

CAMPER'S SECRET

Know what kind of wood you can use before leaving the trailhead on your trip. More and more federal lands are prohibiting campfires, and restrictions are being imposed to preserve what little dead wood is left so it can rot back into the soil to help new trees grow. Never cut a live tree or trim off the branches.

USING KNIVES, AXES, AND SAWS

Most campfire injuries are related to fires and tools such as knives and axes, and safety while using tools must be emphasized at every opportunity. Learn how to fold a knife so that you don't catch your finger. Learn how to use one so that there is never a danger of its slipping and slicing either you or someone near you. Always whittle away from your body. Do not let children play with bowie knives by throwing them against stumps or trees. Not only is it rough on equipment, it is also dangerous and can damage living trees. Emphasize always that knives and axes are tools, not toys.

The use of an ax should be approached with the same respect an animal tamer uses when entering a cage full of tigers. Because of the force required to make an ax do its job, its potential for serious injury is greater than that of almost any other campfire tool.

The illustrations on pages 161-164

THINK SAFE

Probably more camping accidents occur with knives and axes than any other tools, so it is extremely important to set strict rules for the use of each. If your companions—young or old—are not experienced with knives and axes, establish rules early in the outing. Wood-chopping areas should be set aside some distance from the campsite, and you can borrow an idea from the Boy Scouts which involves putting a "fence" around the wood-chopping area with a rope. Children should use knives and axes only with supervision.

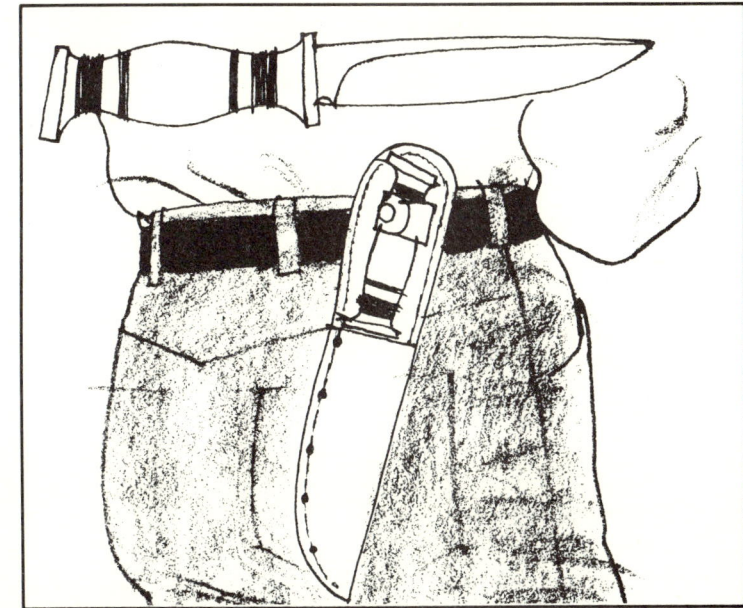

A sheath knife should be placed in your belt in such a way that it won't interfere with your movement or stick you when you sit down.

160   FAMILY CAMPING

Camping knives range from two-blade models to the multiblade-and-tool Swiss army knives.

Always whittle away from your hands and body.

Carry a whetstone to keep knives and axes sharp, and keep your fingers out of the way when sharpening.

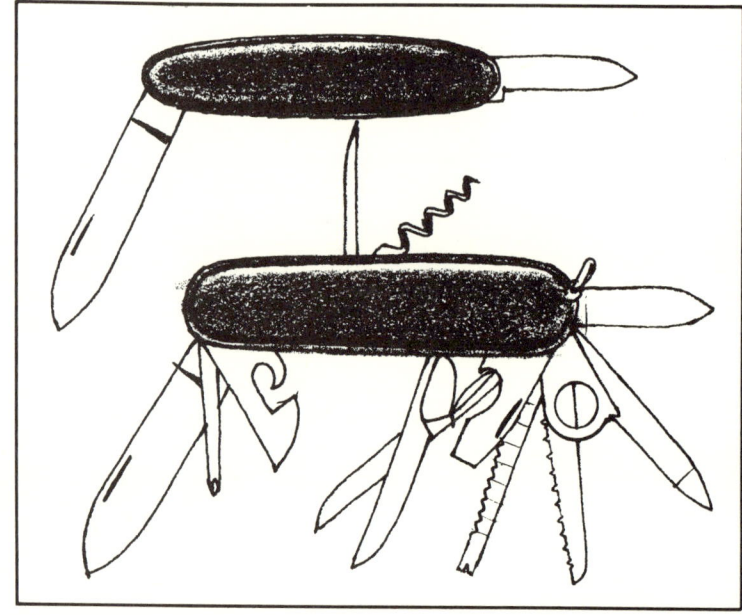

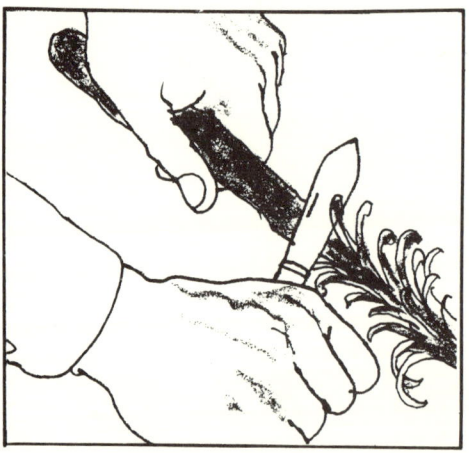

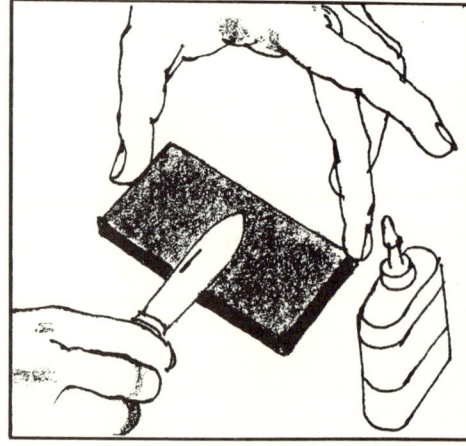

CAMPING ETIQUETTE

The more children know about outdoor lore, the more they will enjoy camping. It is also the perfect place to teach them the basics of outdoor good manners and environmental responsibility. After you have camped at several different sites and the children have seen the evidence of thoughtless campers who came before you—litter, fire-blackened boulders, trees scarred by carved initials or ax marks—they will take more interest in keeping campsites cleaner.

CAMP ACTIVITIES   161

show examples of how to use an ax properly in most camping situations. In all cases, always carry an ax or hatchet in a sturdy sheath.

In many campgrounds you won't really need an ax or hatchet, and a folding saw will be sufficient for all your uses, plus a sharp knife for making shavings, and tinder for starting campfires.

A variety of folding saws are on the market, all considerably lighter than the smallest ax or hatchet. One of the most popular is the Sven saw, which folds into a compact tube that can be tucked away in a pack or in the truck of your car without danger of snagging something.

Other saws are available, including the kind found in many survival kits, which is simply a thin wirelike piece of steel with teeth and with a ring on each end for you to hold.

Most environmental protection

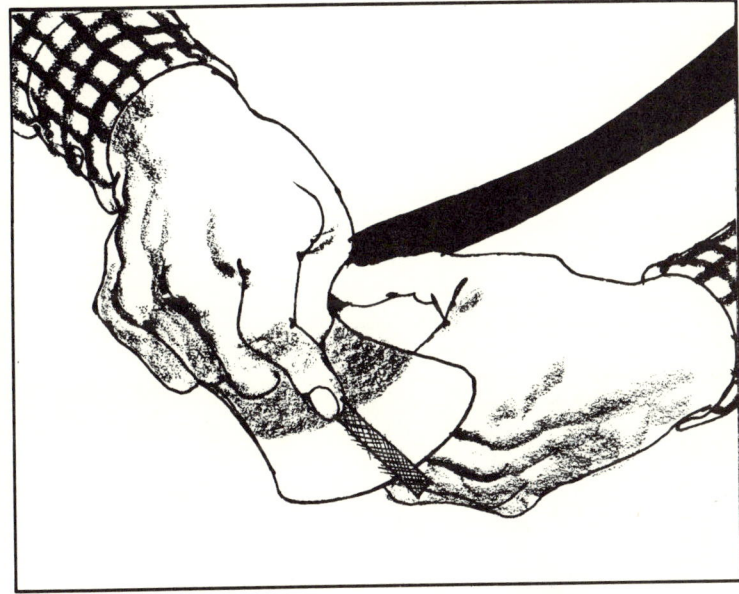

A mill file and double-grit whetstone are needed to sharpen axes and hatchets, the file to knock off the burrs and the heavier whetstone to dress and sharpen the hard steel.

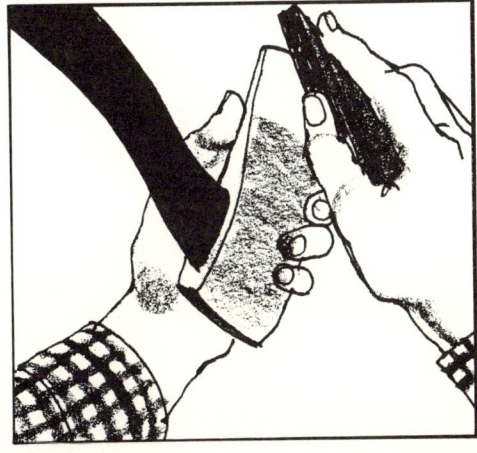

MAKING A HARDWOOD WHISTLE

Many of us remember that great sense of accomplishment when we made our first successful whistle from a hardwood branch. Where tree growth is abundant, you can cut a branch and, usually after a few false attempts, make a one-toned flute.

The branch section should be free of knots or other imperfections. Cut it into about a six-inch length, then gently tap the bark all around until it loosens from the hardwood beneath.

Then cut a notch in the middle, as shown in the illustration, and a notch for the mouth piece. You may not get it right the first time, but after a series of minor adjustments with a sharp knife, you will note that the closer you get to the right cuts, the closer the sound of your blowing through it comes to a whistle.

You can wedge the head of an ax or hatchet between two pairs of stakes to anchor it, placing a piece of firewood between the stakes for more security.

CAMP ACTIVITIES 163

When cutting a stick, always cut at an angle to reduce the distance it will fly. Make the first cut deep enough to hold the wood with the ax; then hold the stick with one hand and swing it downward on a chop block or log to complete the cut. This gives you much more control over the flying wood.

164 FAMILY CAMPING

Use the same technique for splitting. Chop into the end of the stick, leave the ax in, and bring it down on the chop block.

organizations strongly discourage the use of an ax or hatchet by campers. They argue that, as with owning a firearm, you feel compelled to use it whether it is needed or not. This is particularly true with children.

On the other hand, if you check into a Forest Service campground and find the wood supply consists of pieces of wood as thick as a runner's thigh, you will wish you had brought along an ax.

WILDCRAFTING

A pleasant way to combine camping with a chance for school-age children to earn extra income is to camp in areas known for having marketable plants and parasitic growths. In those parts of the country where these things grow in profusion, it is possible for a family not only to recover the costs of camping trips, but also to turn a tidy profit. Some students have paid most of their way through college by "brush-picking" or other forms of wildcrafting.

Here are some examples of forest products for which there is usually a market:

For use in floral arrangements, wreaths, etc:
- evergreen huckleberry
- salal
- sword fern
- scotch broom
- false boxwood
- dwarf Oregon grape
- galax
- leucothoë
- mountain laurel
- mistletoe
- holly
- evergreen boughs

Used by pharmaceutical manufacturers for medicines:
- virginia snakeroot
- cascara bark
- quinine conk

Collecting seed cones for timber companies is another source, but quite specialized.

If you plan to undertake this activity, check first with the local Forest Service office or other governmental agency for information on which forest products have a market value, how to go about getting permission to harvest them, and where to sell them. Some timber companies assign portions of their land to individuals who earn a comfortable income from this line of work.

Most of this activity is in the Pacific Northwest, from northern California to upper British Columbia, and in the Appalachian Mountains.

CAMPER'S SECRET

One chore most children enjoy performing on long trips is making a chart to show which oil companies will honor your credit cards. The names of many oil companies change from state to state, and the various brands under the same parent company are shown on the reverse side of your credit cards. Have children list your credit cards across the top of a sheet of paper, then the different brand names by state beneath. It gives them something useful to do and greatly simplifies your search for a service station.

CHAPTER 10

FIRST AID AND SAFETY

Cuts, scratches, minor burns, insect bites and stings, and other minor injuries that are common at home and in the backyard are to be expected sometimes while camping, too. Be pleasantly surprised if you don't have to open the first-aid kit on a family outing. No matter how careful you are, someone will pick up a hot dish, step on a sliver, get scratched on a sharp rock, or cut with a knife or on the sharp edge of a grill.

Most camping injuries are in this minor category, and you should be prepared for them with a well-stocked first-aid kit. Be sure to take an inventory of the kit's contents before each outing in case someone raided it for a cut at home.

Although serious injuries are unlikely to occur, it is wise to know in advance how to deal with them. Many families, as a matter of course, take the Red Cross first-aid classes offered in most community centers or at work. The Red Cross has made inroads into the public's general medical ignorance with these courses, which are very reasonable in cost and taught by experienced personnel. It will give you a feeling of security both at home and in the outdoors, and this alone makes the time invested well worth it.

Treatment of injuries continues to change and improve. Some of the old ways have been proven to be almost as damaging as the injury itself. For example, we used to be taught to treat frostbite by rubbing it with snow. Now we know that not only should the frostbitten area not be rubbed, it should be warmed instead of chilled. There are many examples of this in first-aid history, so you should be certain your home first-aid manual is up to date.

A BASIC FIRST-AID KIT

No two first-aid kits are alike because each group of campers has its own particular needs, such as prescribed drugs and personal preferences. However, here is a basic list of first-aid equipment that should always be with you when away from home, either on an extended trip by car or public carrier or while camping.

Medicine

Prescription drugs with dosage clearly labeled on the container. Each drug in a separate container. Wrap dosage instructions in clear plastic tape to prevent smearing or erasing.

- **aspirin**—at least 12 tablets, more if you expect to need them
- **salt tablets**—24, to prevent heat cramps
- **antacid**—at least 6 tablets, in case the cook blunders
- **antihistamine**—at least 6 tablets, for insect bites, stings, colds, etc.

CAMPER'S SECRET

When you need a cold compress to reduce swelling, don't forget to look in your ice chest. It sounds insultingly simple, but when you're in pain and the others are worried, it's easy to forget the obvious.

Bandages

- **bandage strips**—a dozen one-inch strips for minor lacerations
- **butterfly bandages**—6 or more, for closing lacerations
- **Carlisle battle dressing**—one or two, for large wounds
- **moleskin**—a 6-inch-square sheet or more, for blisters
- **sterile gauze pads**—for large wounds
- **tape**—2-inch roll, for holding bandages on or for sprains
- **triangle bandage**—for supporting an injured arm or covering a large dressing
- **elastic bandage**—one 3-inch size; know how to use it properly

Tools

- **needle**, to remove splinters, etc.
- **razor** (disposable is fine)—for shaving hairy areas before taping
- **safety pins**—3 or 4 for mending clothing, holding bandages, etc.
- **oral fever thermometer**

Liquids

- **tincture of benzoin**—1-oz. bottle, to help hold tape to skin
- **antibacterial soap**—1-oz. bottle, for mild antiseptic cleansing of cuts

Suggested Additions

coins for telephone calls
small first-aid book
thread for clothing repairs
sunscreen to prevent sunburn
anesthetic lotion for use if sunburn prevention fails
Caladryl lip balm

You should also carry a snake-bite kit when traveling in areas where poisonous snakes are common—but you must know how to use the kits properly and be prepared to do so. Most kits are equipped with a small knife for lacerating the bitten area and a small suction

FIRST AID AND SAFETY

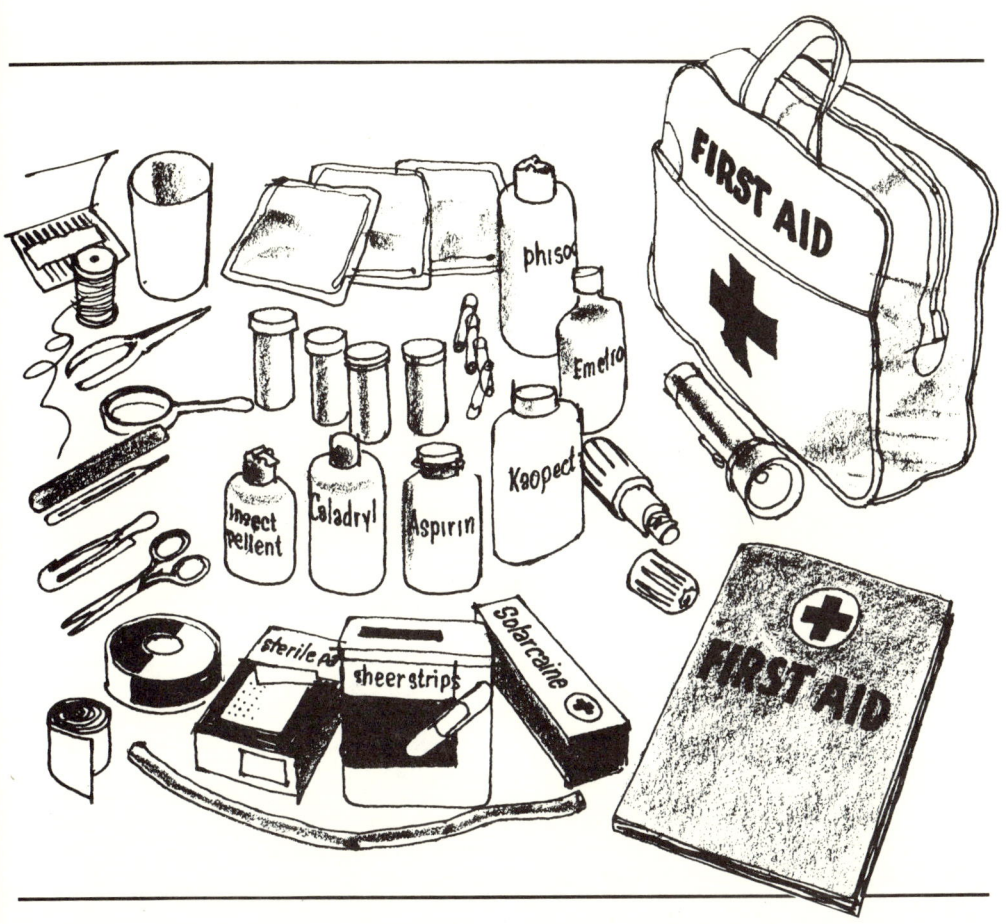

device for removing as much of the venom as possible. Some also have antivenin to counteract the poison. Check with your family physician for the latest information on snake-bite treatment. Very few people are bitten by snakes each year, and deaths are very uncommon, except from bites of the coral snakes in the southeastern United States. Obviously, as with all camping accidents, taking preventive measures is the wisest course of action; everyone should be trained to watch carefully for snakes.

Burns: For minor burns, immerse the burned area in cold water to relieve the pain, then carefully wash the area with a liquid soap using sterile cotton soaked in water that has been boiled and cooled. Cover the burn with a layer of sterile dressing, preferably coated with petroleum jelly so it won't stick to the burned skin. Then cover this with a larger, thicker bandage that applies only a moderate amount of pressure to the injured area. Leave for approximately a week without removing to avoid the possibility of infection.

In cases of minor first-degree burns, bandaging isn't necessary except to prevent further injury to the area. Burn ointments, similar to that recommended for sunburn, can be applied to superficial, cooking-type burns.

If the burn covers a large area, however, or destroys the skin, take the victim to a doctor or hospital to avoid further damage. Keep all clothing, jewelry, and other foreign matter from touching the skin and keep the victim as immobile as possible so additional damage to the tissue won't occur.

Heat exhaustion and heat stroke: The best treatment obviously is to take preventive measures. An adequate intake of salt and fluids will prevent most from occurring.

Heat exhaustion is caused by prolonged physical activity in a hot climate and is caused by the blood vessels in the skin becoming so dilated that they rob the supply to the brain and other vital organs. The result is similar to fainting and usually isn't serious. The victim will feel faint and often has a rapid heart beat. Sometimes the victim will also feel nauseated and have a headache. Rest, along with an intake of salt and fluids, is the best treatment.

Heat stroke (or sunstroke) is in many ways the opposite of hypothermia and can be fatal if not treated quickly and properly. The body must be cooled as rapidly as possible, either by immersion in tepid—not cold—water, application of wet towels or blankets over the entire body or, best of all, with towels soaked in alcohol.

The injury may come quickly. The victim will be normal one minute and in a very short time become confused, irrational, and uncoordinated. At this stage,

WATER SAFETY

It is always surprising to find how many people never learn to swim or bother to learn or have their children learn the rudiments of water safety. Several thousand people drown each year, and most of those are in still water rather than swift rivers or far out to sea. Most occur within sight of the beach, and most could be prevented if more people were at ease around the water.

Water safety should be an important part of your outdoor experience, not only the ability to swim but also the ability to stay afloat in water for long periods.

One way to do this is testing yourself at the local swimming pool or lake. Try the lifesaving techniques, such as using your clothing for life preservers. Most cloth is airtight when it is wet, and pants or long slacks make good life preservers. Tie the ends of the legs together and, while treading water, swing the pants over your head with the opening down. This will at least partially fill the legs with air, and you can inflate them the rest of the way by blowing into them. By keeping the opening below the surface, you'll be able to float for several minutes, or longer if the material is closely woven.

No matter how well you swim, you should never be out in a boat without a life vest on, and children should grow up taking these vests for granted.

Your local Red Cross chapter will offer lifesaving classes that include the accepted methods of artificial respiration for drowning victims. The odds are against your ever having to use this knowledge, but the peace of mind you'll have from the knowledge is well worth the small amount of time and the modest expense involved.

the victim will usually have a body temperature of 105 degrees Fahrenheit or higher and no sweating at all.

The treatment must begin immediately. Sunstroke is one of the most dangerous accidents because death or brain damage can occur if the body's temperature isn't brought down to at least 102 degrees rapidly. The victim will not recover for some time, and the temperature may fluctuate up and down for a few days afterward.

OUTDOOR PESTS

Pesky critters that buzz through the air or crawl about are a natural part of outdoor experiences, from backyard picnics to wilderness treks. Perhaps as much as fear of discomfort, the dread of these pests makes many novices nervous about camping. Some people have an almost irrational fear of harmless pests, and there apparently is little that can be done to remove the fear, other than wearing it away by finding places to camp with a minimum of pests, and spending as little time as possible talking about it. Like fear of the dark, talking about it can only make it worse; treating it like the normal thing it is will gradually erode fear into an easy acceptance.

These outdoor pests come in almost as many varieties as there are campers. Rather than go through the whole encyclopedia of things that bite, sting, or stink, let's consider the most common pests and how to coexist with them.

Ticks live mainly in hardwood forests and open range where they have cattle available. They are seldom in the high country above timberline, and they do not like cold, damp climates. If you are camping in an area where ticks may be found, be sure to check your body at least once a day, especially around your boot tops, belt line, armpits, crotch, beneath folds in your skin, etc.

If you wear high-top boots, tuck your pants legs in them if possible or into your socks, then spray your pants legs and socks with an insect repellent. This may not keep all of them off, but it will help.

When you find a tick on your body, remove it immediately and carefully. If it has only its head buried in your skin, gently ease it out with your fingers or tweezers. If it doesn't dislodge easily, apply a lighted match or cigarette to its body. Then kill it.

If the tick has burrowed into your skin deeply, apply heat to it. Sometimes a bit of alcohol or stove fuel (white gas or kerosene) will remove it.

Should you accidentally leave the head burrowed in your skin, remove it

---

CAMPER'S SECRET

Some outdoor pests are more of an irritant than a major problem, such as the slugs and big snails common in some areas of the country. They are disgusting to touch, leave a trail of slime behind them as they inch along and, if they are accidentally stepped on, leave a mess on your shoes and the ground. They can be killed by sprinkling salt on them, but you'd have to carry bags of salt to build a salt fence around the campsite. No final solution is known, although some people put out saucers of beer to attract them to their death by drowning.

TICKS

Rocky Mountain wood tick

American dog tick

Lone Star Tick

immediately with a bit of minor surgery. Sterilize a sharp knife and make a small incision, then treat the area with an antiseptic. In all cases, an antiseptic should be used.

A few diseases are occasionally transmitted by ticks, even though the incidence of Rocky Mountain tick fever and other afflictions is on the decrease. So take no chances.

CAMPER'S SECRET

One method of keeping flies and stinging insects such as hornets and bees away from the camping area is to pour a small amount of syrup or honey in a dish and place it well away from the campsite. It won't attract all of them, but it will help.

Spiders: Thanks to a host of horror movies, spiders elicit a stronger fear among us than they actually rate. Although all spiders are venomous, only a tiny fraction of the species have teeth strong or long enough to penetrate human skin. And of these, only two are likely to be a problem for humans. Fortunately, few live beyond urban areas because their food supply is the flies, gnats, and roaches that also depend on man's garbage for food.

The black widow is the most dangerous, although almost every fatality from one's bite is among very small children or the elderly; individuals in good health almost always recover. Black widow spiders are about one-half inch long with an almost spherical body. The underside has a red-orange hourglass marking. After a person is bitten by a black widow, severe pain usually starts within an hour in the area of the bite, then spreads to the rest of the body. The victim is usually quite ill with abdominal pain and spasms, and often shock.

Antivenin is available but should be administered by a physician. Most treatment, such as hot baths, is directed at relieving the muscle spasms. But the victim should be taken to a doctor or hospital as soon as possible. The pain and prostration caused by the bite usually disappear in two to four days, but the victim will usually be weak for a few weeks afterward.

The other dangerous spider is the brown spider, called the brown recluse or violin spider. It isn't often fatal, except to very small children. The brown recluse is slightly smaller than the black widow and has an oval body with a dark violin-shaped spot on the front portion of its body. Its bite causes a severe inflammation in a small area around the bite. Over the next two weeks, the resulting blister usually ruptures and the skin turns dark and falls away, leaving a sore that usually heals with some scarring.

The most feared spider, because of its ghastly appearance, is the large tarantula, native to the southwest desert and parts of the South and Midwest. Its bite is no more damaging than a bee sting; it just looks awful.

When walking through underbrush, where spiders build their webs between branches of trees or bushes, carry a walking staff or simply a stick to beat the webs away before you get to them.

When picking up a stick or rock, first roll it over with your toe or a stick.

Carry insect spray designed to kill spiders and use it on areas with lots of dead wood or rocks near your campsite.

Always check your sleeping bags and clothing for spiders or insects before using them.

Scorpions can inflict a bad sting. Since they are primarily nocturnal, especially in desert areas, you probably won't see them unless they find their way into your clothing or personal effects. So always shake out clothing, bedding, and towels before use in areas where scorpions are common. Be careful when lifting rocks or old boards, where they like to hide.

Bears: Although they hardly qualify as pests, owing to their size, many campers fear bears without knowing the difference between species. Except for Montana and Wyoming, particularly in Glacier National Park and Yosemite National Park, the grizzly bear is not common in the contiguous United States. They are common, however, throughout western Canada and Alaska and should be treated with great respect. The major safety precaution with grizzlies is guarding your food supply. Park rangers have been trying for decades to educate campers not to leave food and garbage around so they won't attract bears to their camping area. If you are camping in an area known to have grizzlies, keep your food suspended at least ten feet above the ground from a rope hanging from a tree branch. Do not keep it in your tent. Do not sleep near the campground garbage dump.

For the rest of North America, the smaller black bears are even more frightened of us than we are of them. Their major contribution to camping problems is that they, too, are given to raiding the food larder. Thus, the necessity of carrying a sturdy container, preferably plastic, in which all your food will fit and a rope to sling over a high tree branch.

The prime rule of safety with bears is to be certain they know you are in the area. Many hikers carry small bells on their packs that aren't loud enough to irritate other hikers, but let the bear know they're around. Some people rattle a small stone in a tin can as an alert for bears that might be feeding in a nearby thicket. Others say that a thin willow whip will create a noise that frightens bears away.

Whatever you use, it is important that bears not be surprised by your presence, and particularly sows with cubs cavorting around them. Sow bears will take drastic measures to protect their cubs, and if you find yourself near a cub, depart immediately. Cubs are as much fun to watch as a pup or kitten, but their mothers can become quite cranky if they fear their offspring are in danger.

Mosquitoes are the most common of all pests and are found almost everywhere you are likely to camp. Insect repellent, netting on your tent, and long pants and long-sleeved shirts are the best insurance against bites. Mosquito coils, rings of a very slow-smoldering and evil-smelling chemical, will clear them out of cabins and keep them out as long as the coil burns.

Chiggers and noseeums are among the most irritating of all pests because they are so difficult to see, and their bites cause itching. As with ticks and mosquitoes, the best approach is prevention: apply insect repellent to your boots and pants legs. Obviously, camping in shorts gives them a bigger banquet table.

Ants seem to be almost everywhere, especially where food is kept. Other than the irritating prospect of finding them in your cup or in your bedding, they're of little consequence. You can usually keep them away by hanging food from tree branches. Check the campsite before settling down for a long stay in case there is a hill of ants nearby. If so, let them have the area.

Bees and hornets: Virtually everyone has been stung sometime by a bee or hornet. Although painful, the stings can be treated easily with a paste of baking soda and water. Some people who become allergic to their stings should be treated with antihistamines carried for those purposes.

A word about insect repellents: Some people swear by massive doses of various vitamins as the best insect repellent; others say eating garlic by the pound will keep them off your body—and your companions at their distance as well. There apparently is no single insect repellent that works best for everyone, and you'll have to experiment with two or three of the major brands before you find which is best for you. They come in spray cans, squeeze bottles, lotions, and sticks, and have brand names such as Off, Cutters, Jungle Juice, and Mirehol.

Warning: Do not let insect repellent spray hit your tent wall or any other synthetic material. Some varieties will weaken or destroy these fabrics. Read the manufacturers' warnings for each tent, and ask outdoor-equipment stores about other items, such as rain gear.

Skunks: When you come across a skunk in camp or on the trail, let it have the right of way. It won't spray you with its pungent and clinging urine unless it is frightened or cornered. With justification, skunks are not very frightened of people, but they don't lurk around camp very long if they know people are wandering around as well. If one does spray you, about the only thing you can do is bury your clothing and take a series of baths to remove the stubborn odor. Vinegar is one agent that will eventually remove the odor; tomato juice is another.

Porcupine: These slow-witted camp robbers cause no trouble if you leave them alone. They usually come into camp in search of salt, and that salt may be on your clothes or backpacking equipment. They can be frightened away and are harmless unless you make the mistake of touching them. Dogs are the biggest problem with porcupines because they never know enough to leave them alone, and a dog with a nose full of porcupine quills is in agony until they are removed. Usually they have to be removed by a veterinarian.

TREATING PROBLEMS

Following are methods of treating some of the most serious wounds.

Stretchers can be made of a variety of things at hand in case a member of your group must be carried for some distance. One of the most simple consists of two poles, made from saplings cut and with the branches trimmed off, and two or three jackets or parkas.

The parkas should be zipped up or snapped, however they are closed, and the poles inserted through the arms and the bottom of the parka. Two of these will usually suffice for an average-size adult, but more can be added if necessary.

If you are carrying the victim over rough terrain, you may need to lash the victim in, particularly if he or she is unconscious or very weak.

Make the victim as comfortable as possible by using other articles of clothing or sleeping bags as padding.

Shock is usually caused by a sudden reduction in the volume of the victim's blood, either as a result of severe bleeding or when the blood serum is rushed by the body's healing mechanism to a severely burned area. It can

CAMPER'S SECRET

It is hardly a secret that you should always carry some emergency provisions and tools in your car. One of the most important is a shovel, either a short-handled and sharp-pointed one or a folding shovel so popular in the military services. This can be used around the campground for digging latrines, for fire safety, and for similar uses. You should also carry emergency flares that are easy to ignite, and a block of wood or something similar to block your wheels in case you have to change a tire or do some other automotive surgery. Another convenient tool is a small winch that can be used to extract your car when it is stuck in the sand or mud. A portable winch with strong cable or cord plus a shovel will get you out of most situations provided you have something strong, like a tree or another vehicle, to winch toward, and also provided you don't dig the car in too deeply before giving up.

also result from dehydration when the body's fluids are lost through vomiting or diarrhea.

When one of these causes occurs, the arteries constrict to divert the blood supply to the vital organs, and the heart begins pumping faster to circulate the blood that remains or to help in the healing process.

A victim of shock is pale, with skin turning cool, first in the extremities and later the trunk. As shock becomes more serious, the patient will sweat and complain of being cold and frequently thirsty. The pulse is often rapid and breathing fast and shallow as the blood pressure lowers.

It is best to treat potential causes of shock quickly to prevent shock from occurring, or at least to minimize it. Bleeding should be controlled immediately, because it is the loss of blood, not the initial injury, that causes shock. In the case of burns, fluids should be given to minimize the loss of other fluids to the injured area.

The victim should be made to lie down and keep both feet elevated about twelve inches above the level of the head. Hot water bottles, heated stones, or body-to-body contact beneath plenty of cover should be administered to help raise the body temperature. Keep the victim prone until treatment has been completed and he or she appears to be in stable condition.

In all cases—even when someone has suffered only mild shock—he or she should be taken to a doctor as soon as possible after being stabilized.

Fractures are usually quite easy to diagnose, although they sometimes can be confused with severe sprains in joints. In either case, the victim should be immobilized and taken to a doctor for treatment.

If the fracture isn't compound (i.e., if the bone isn't protruding through the skin), the area of the break will be painful to the touch; swelling and discoloration will set in soon, and a lump or crook will show in the bone.

CAMPER'S SECRET

A small vial of lighter fluid can be carried and used to remove adhesive tape without irritating the skin.

Even if these symptoms are not evident—though they usually are—the victim should lie down and you should make a splint to cover the injured area so more damage won't occur. Do not make an attempt to set the broken bone; that is the job of a doctor. But do make every attempt to prevent more damage through the construction of a splint.

Splints can be made from a variety of materials at hand, from small saplings cut for the purpose to pieces of boards or tent poles. For them to work properly in immobilizing the fractured area, splints must anchor the limb both above and below the fracture.

The splint should be padded between the rigid surface and the places it touches the body. Wrap it firmly at once, though not so tight that it interferes with blood circulation.

If the fracture is below the knee, the victim can move on the uninjured leg with assistance, but this should not be permitted unless necessary. Fractures of the knee and above require that the victim be carried on a stretcher, as do fractures of the hip and pelvis or any spinal injury.

Arm, wrist, and shoulder fractures and dislocations should be immobilized, also. Fractures of the arm should be splinted and the arm immobilized against the body trunk with a sling.

Severe cuts: It has been only a short time since doctors have recommended that we forget about learning the so-called pressure points throughout the body where you can apply pressure to stem the blood flow through arteries to injured areas. Now even tourniquets have fallen into disfavor because, like pressure points, they can lead to more damage if not used properly.

Pressure directly to the wound is the only safe and effective means of stemming blood loss. This collapses the severed blood vessels so that clotting can occur.

Arterial bleeding, identified by bleeding in spurts as opposed to steady bleeding from veins, is the most serious kind and can also be controlled by direct pressure, although it sometimes takes longer. If the bleeding doesn't stop after several minutes of holding gauze bandages on the wound, wrap the injured spot tightly, but not so tightly that the blood supply is shut off to the rest of the limb. If the skin beyond the bandage turns dark, or if the patient complains of a tingling or numbing sensation, the bandage is too tight.

POISONOUS PLANTS

Poison oak

Before taking your first camping trip, be certain everyone in your group can identify the major poisonous plants in the camping area.

Poison oak and poison ivy are the most common. They are almost identical, except that ivy is a vine and the oak variety is a bush. Each has a cluster of three leaves and clusters of white berries.

Poison sumac closely resembles the more common staghorn sumac except that its leaves have smooth instead of jagged edges and its berries are white or gray instead of red.

Nettles, which can be boiled and eaten, are another irritating plant that can cause a rash, especially if you scratch the area they have touched.

Scratching should be avoided because it can cause infections or scarring. Calamine lotion helps relieve the itching, and a prescription drug, Syn-

Poison ivy

Poison sumac

alar, or its weaker cortisone relatives such as the creams and lotions sold over-the-counter will help.

If the rash is over a large area, relief can be gained by making a solution of saltwater (two teaspoons of salt per quart of water) and applyng a compress soaked in this water several times a day.

No area of America is totally free of outdoor creatures and plants that can be an irritant to people, and this includes your own backyard. Yet millions of campers take these minor disadvantages for granted and, by recognizing them and taking normal precautions, do not let them become problems.

HYPOTHERMIA

Hypothermia has only recently entered our vocabulary, replacing exposure as a threat to outdoors enthusiasts. Fortunately, since the problem has been studied thoroughly and received much publicity, fewer and fewer deaths have occurred from this major problem.

Essentially, hypothermia is the loss of body heat. When body temperature drops, more than three degrees below the usual 98.6 degrees Fahrenheit, certain changes occur, and the lower the body temperature drops, the worse these changes become. Unless reversed, hypothermia can lead to death.

However, as stated above, most outdoors people have been indoctrinated against this, and in most cases it is no more complicated than telling your children to wear warm coats when they go outside in the winter. The problems come when people go on hikes or other outings on a warm, sunny day, and then a cold rainstorm comes and they are not prepared for the cold and wetness.

Thus, the importance of having everyone in a camping group always carry rain gear, a warm sweater or jacket, and the other Ten Essentials (see CHAPTER 3) when leaving the campsite. If you are camping in higher elevations, the weather can turn cold faster than at or near sea level. Many, many campers have never given a thought to hypothermia because they are so well equipped. To repeat, always carrying the proper clothing is the best prevention.

While prevention is relatively simple, except in the case of campers who might be stubborn to the point of stupidity (and you should find someone else to camp with in this case), a great deal of research is still being conducted to test various ways of speeding the recovery of hypothermia victims.

Basically, hypothermia is broken down into several stages. The first stage, which usually sends people in search of a warmer coat or wool socks, is shivering.

The next stage is violent, uncontrol-

FIRST AID AND SAFETY

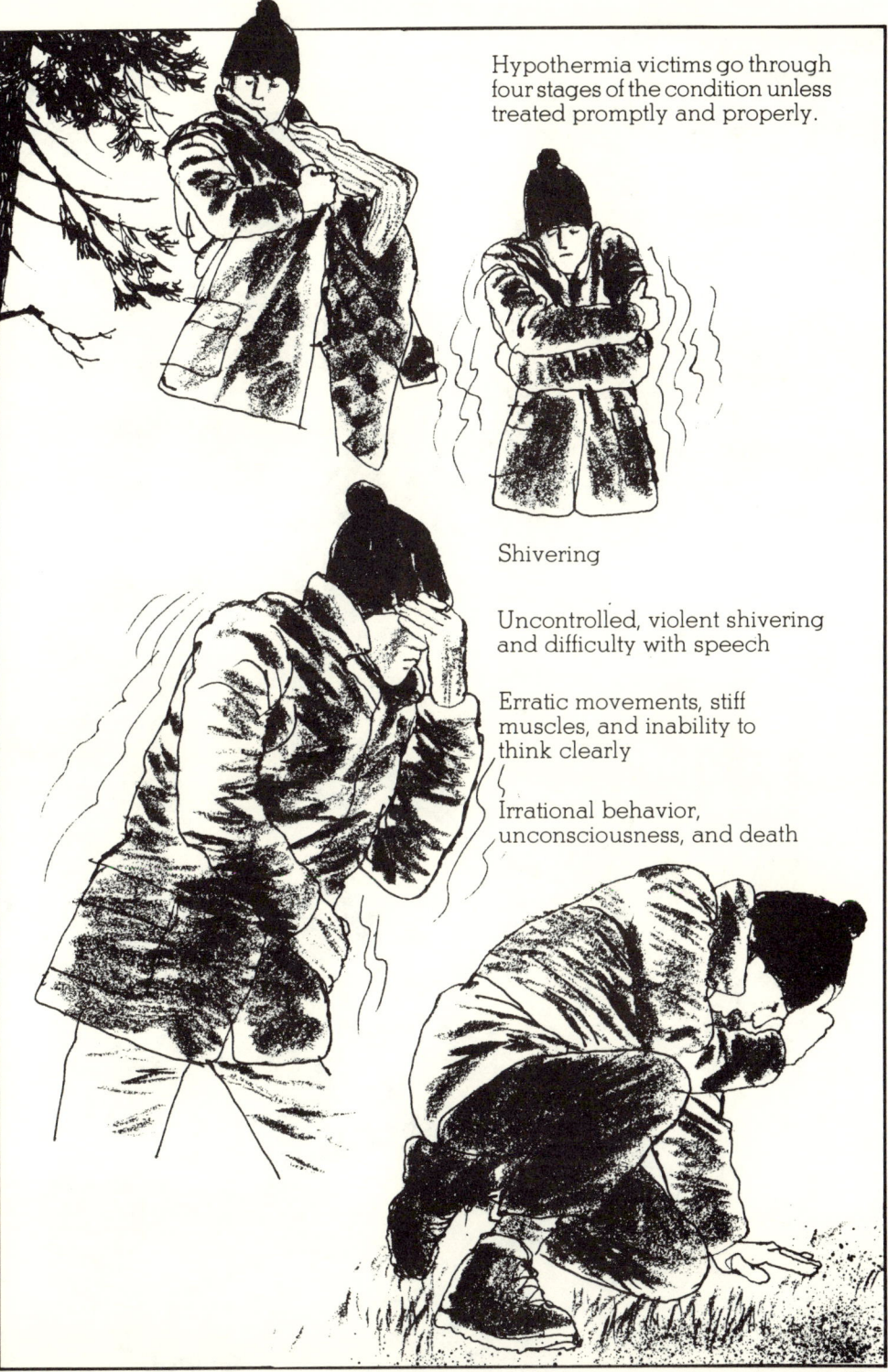

Hypothermia victims go through four stages of the condition unless treated promptly and properly.

Shivering

Uncontrolled, violent shivering and difficulty with speech

Erratic movements, stiff muscles, and inability to think clearly

Irrational behavior, unconsciousness, and death

180 FAMILY CAMPING

It is essential to treat the hypothermia victim quickly.

Get the victim warm and into dry clothing

Stoke up a big fire.

Walk the victim around to force his or her body to rebuild its own heat sources.

Feed the victim hot drinks.

Once the danger is past, keep the victim warm and comfortable.

FIRST AID AND SAFETY

Wind Chill Chart

COOLING POWER OF WIND EXPRESSED AS "EQUIVALENT CHILL TEMPERATURE"

WIND SPEED MPH	TEMPERATURE (°F)																				
CALM	40	35	30	25	20	15	10	5	0	-5	-10	-15	-20	-25	-30	-35	-40	-45	-50	-55	-60
	EQUIVALENT CHILL TEMPERATURE																				
5	35	30	25	20	15	10	5	0	-5	-10	-15	-20	-25	-30	-35	-40	-45	-50	-55	-65	-70
10	30	20	15	10	5	0	-10	-15	-20	-25	-35	-40	-45	-50	-60	-65	-70	-75	-80	-90	-95
15	25	15	10	0	-5	-10	-20	-25	-30	-40	-45	-50	-60	-65	-70	-80	-85	-90	-100	-105	-110
20	20	10	5	0	-10	-15	-25	-30	-35	-45	-50	-60	-65	-75	-80	-85	-95	-100	-110	-115	-120
25	15	10	0	-5	-15	-20	-30	-35	-45	-50	-60	-65	-75	-80	-90	-95	-105	-110	-120	-125	-135
30	10	5	0	-10	-20	-25	-30	-40	-50	-55	-65	-70	-80	-85	-95	-100	-110	-115	-125	-130	-140
35	10	5	-5	-10	-20	-30	-35	-40	-50	-60	-65	-75	-80	-90	-100	-105	-115	-120	-130	-135	-145
40	10	0	-5	-15	-20	-30	-35	-45	-55	-60	-70	-75	-85	-95	-100	-110	-115	-125	-130	-140	-150

WINDS ABOVE 40 HAVE LITTLE ADDITIONAL EFFECT.

LITTLE DANGER	INCREASING DANGER (Flesh may freeze within 1 min.)	GREAT DANGER (Flesh may freeze within 30 seconds)

DANGER OF FREEZING EXPOSED FLESH FOR PROPERLY CLOTHED PERSONS

Source: National Weather Service, U.S. Dept. of Commerce

Summer* Weather Chart

Regions	Normal Monthly Precip. (in.)	Normal Daily Max. Temp.	Normal Daily Min. Temp.	Normal Daily Mean Temp.	Average % of Poss. Sunshine
Pacific N.W.					
Seattle	1.08	73.8	53.7	63.8	62
Spokane	.58	81.9	54.0	68.0	77
California					
Bay Area	.03	71.6	54.3	63.0	65
Los Angeles	.02	75.8	63.2	69.5	83
Interior	.05	91.3	56.9	74.1	96
Colorado					
Denver	1.29	85.8	57.4	71.6	72
MN/WI					
Minneapolis/ St. Paul	3.05	80.8	59.6	70.2	67
IL/MI					
Chicago	2.73	82.3	59.9	71.1	68
Detroit	3.04	81.6	62.1	71.9	65
N. Atlantic States					
Boston	3.46	79.3	63.3	71.3	67
Mid-Atlantic States					
Wash. D.C.	4.67	86.6	67.6	77.1	63

*Figures are for August, normally the warmest, sunniest month. Source: U.S. National Oceanic and Atmospheric Administration, as reported in "Statistical Abstract of the United States," U.S. Dept. of Commerce, Bureau of the Census. Temperatures recorded at airports, in Fahrenheit degrees.

lable shivering and difficulty in speech.

Then, unless checked, the shivering decreases and muscles become stiff. The victim makes erratic movements and can't think clearly.

If the condition is allowed to continue, the victim becomes irrational and loses contact with reality. Unconsciousness is the next step. Death is the last.

As with any response to one's environment, hypothermia affects each individual differently. One camper may be shivering miserably while the other, dressed in identical clothing, will be standing around with coat unbuttoned, complaining that he or she is too warm. It isn't a sign of weakness to get cold easily, it is simply due to the physiological differences among us.

It is essential to act quickly. Watch other members of your group for the first signs of shivering and emothional withdrawal. Get them into dry, warm clothing, feed them hot drinks and high-energy foods, such as chocolate energy bars, and make them walk. Stoke up the fire. And remember that this is no time to tease or ridicule the sufferer.

Some researchers on the subject say it is best to keep the chilled person moving about, building up body heat with warm drinks (nonalcoholic) and energy-producing foods, and making the body restore its own internal heat. Most researchers say this is the most import part of the recovery process: to help the body restore its own heat, by exercises, warm liquids, and high-energy foods.

Barring dunkings in cold water, few campers will suffer from the more advanced stages of hypothermia if they follow the basic rules of dressing properly for the weather conditions.

CHAPTER 11

STRETCHING THE SEASON

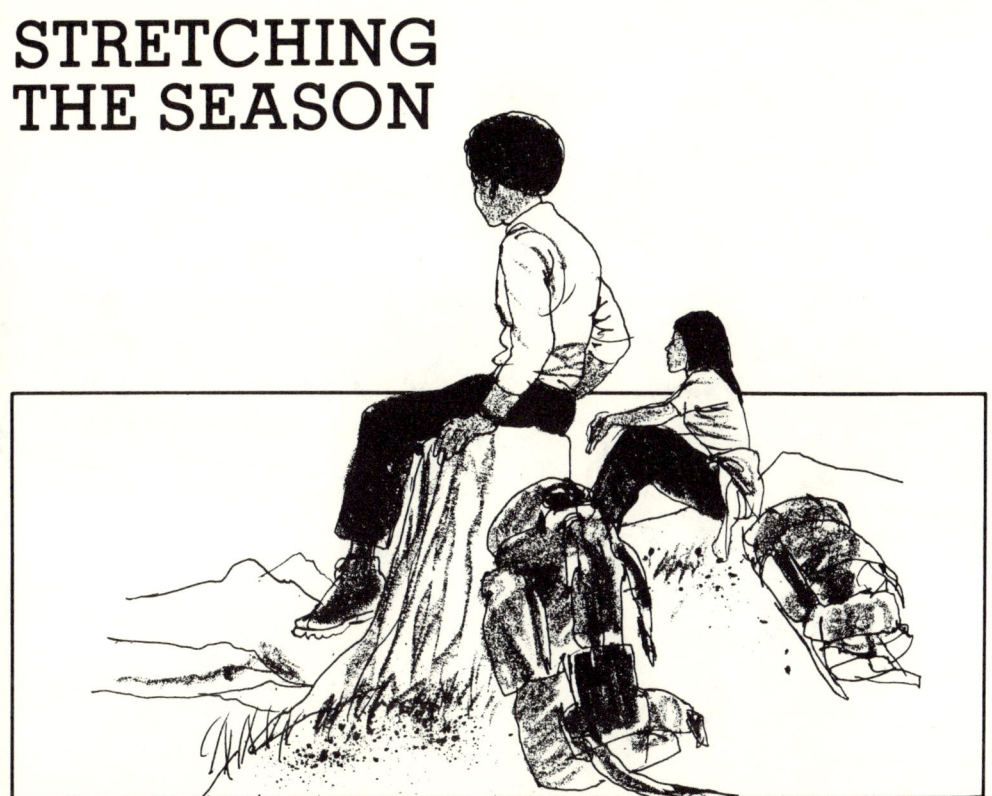

After you have been camping a few times during the summer months and find that it isn't nearly as primitive or as uncomfortable as you feared, you will also discover that with only a modest addition to your equipment list, and a few precautions, you can extend your camping season by several months.

One definite bonus is that before June 15 and usually after September 15 of each year, the camping sites that were crowded during the summer are virtually vacant. School activities keep most families home on weekends nine months of the year, and the majority of campers have put away their equipment until the summer months return.

Winter camping will be discussed later, but for spring and fall camping, with little prospect of snow, your summer equipment of tents and sleeping bags will be sufficient. Following are suggestions on how to stretch the season for your basic equipment.

Sleeping bags: Most are rated to about freezing temperature, or 30 degrees. You can add another ten to twenty degrees by wearing dry woolen clothing to bed, by adding a wool blanket over the bag, or by wearing waffle-weave underwear at night. Some bags can be extended even more by inserting a lightweight goose-down liner or one made of flannel. When you buy your sleeping bag, ask about these options so you won't have to buy an additional bag for colder camping.

Clothing: For spring and fall camping, an additional layer is usually sufficient. The days are usually warm, but the nights have to be taken into consideration. Sometimes this is only a matter of wearing a heavier wool shirt and pants or a goose-down jacket or parka.

Footwear: Your shoes or boots should be leather instead of tennis or running shoes, and you should be able to wear two pairs of socks in them. Down booties are excellent for sleeping and tent wear.

Head wear: A cap or insulated hood should be worn. Sometimes your lightweight parka, essentially a wind parka, will not have the insulation in the hood. Thus, a cap that covers your ears should be worn. Some people do not like hoods unless they are absolutely necessary, so a wool watch cap or balaclava is a necessity.

Gloves or mittens: You might need a pair of each, with liners. Many ski gloves come with removable liners that reflect heat. Mittens can be leather with wool liners or down-insulated with lightweight liners.

Rain protection: Rain is usually more common during the spring and fall than midsummer. Although rain gear should always be taken along on camping trips, it is especially essential when stretching the season. In addition to a raincoat or poncho, rain pants should also be worn. True, no totally satisfactory

rain gear has yet been developed. It is all bulky, tends to overheat when you exert yourself, and makes a lot of noise when walking. But the alternatives of cold and wet discomfort make it essential. A dining fly is also excellent protection against the sun and rain.

Since you must expect rain (and you can be pleasantly surprised if it doesn't come), the best campsites to choose are those above the level of nearby streams, with a slight slope to them so you can avoid seeing your campsite turn into a mudhole.

Food: You will be burning, and needing, more body fuel to keep warm, so your diet should be heavier in carbohydrates and hot drinks. Additional meats such as jerky should be added to your larder, and additional packets of hot chocolate are useful for both heat and energy.

Without getting too technical or scientific, this increased need for fats and carbohydrates means you should stock up on items that many reducing diets would ban: hash-browns, pancakes and syrup, cookies, bacon bars, jerky, and so forth. Gorp with an ample supply of chocolate candy and peanuts is another good source of carbohydrates (the

> **CAMPER'S SECRET**
>
> Spare mitten liners can be made of old wool socks with thumb slots sewn in.

chocolate and raisins) and fats (the peanuts).

Tents: Since few tents are designed to hold heat inside, you can't depend on them for comfort—only for shelter and privacy. You'll have to depend on your clothing and sleeping bag for extra warmth. You can improve your tent so that it isn't quite so crowded by carrying along an eight-by-ten-foot, or slighly smaller, waterproof tarp. You can construct an awning over the tent entrance, or a simple A-frame extension of the tent, and store your gear in the shelter to expand the roominess of your tent.

The same tarp can be used as a cover for the kitchen area by stretching it overhead as an awning or as a wind break.

It is essential that the tent be kept clean, especially if it is cold and raining, and this A-frame tarp over the entrance will allow you to remove your wet rain gear and muddy boots before entering the tent.

Each of the major basic tent designs discussed in CHAPTER 4 have modifications for use in cold-weather activities such as cross-country skiing and mountaineering. Some cold-weather variations available include the following:

Vestibule: This is an extension over the entrance to give extra protection from the wind, rain, and snow. Some models offer it simply as an extension of the rain fly, others as a part of the entire tent. You can store extra gear in the vestibule, or use it for the camp kitchen since the danger of damage from fire or spilled food is less.

Tunnel entrance: Because all zippers are susceptible to freezing, many cold-weather tents have a tunnel entrance with a drawstring at the end. Some have a drawstring at the outer end and a zipper over the inside entrance. Both designs offer a safe entrance in bad weather.

Frost liner: This is a detachable inner wall, often made of cotton, that collects the body moisture that would otherwise condense and form ice on the tent wall and ceiling. The liner can be detached and the ice shaken off at regular intervals. For ease of removal, most are held in place by ties rather than straps or snaps.

Cook hole: Some tents designed for hard use have a zippered hole in the floor that can be removed so that cooking inside presents less of a danger. Spilled fuel or boiling food goes onto the ground or snow beneath instead of the tent floor. As with other modifications, this appears in tents designed for hard use and is not a standard feature of all tents.

WINTER CAMPING

As you become more experienced in the woods, and have stretched the camping season closer and closer to both ends of winter, you probably will become intrigued with the prospect of camping in the dead of winter. The rules of keeping infants happy apply to winter camping—make sure you're dry, warm, and well fed. All gear additions refer back to these three basics.

Generally speaking, most of your summer camping equipment can be adapted, or added to, for winter camping. However, you will want to invest in special winter tents, heavier sleeping bags with a lower comfort range, and special clothing if you plan to make it a regular sport.

Another new factor to contend with

STRETCHING THE SEASON

is transportation. You won't want to "post-hole" your way through deep snow back into the forest, so you will most likely take up winter camping in connection with active winter sports such as cross-country skiing or snowshoeing.

The site you select for your camp should be on a flat clearing with protection from the wind. A good source of water should be nearby. Melting snow for camping water supplies is a tedious process and involves the use of more stove fuel than is necessary.

Avoid camping on hillsides, or at the foot of a bank or hill, because of the danger of avalanches or small snowslides.

If possible, avoid camping beneath evergreen trees because they often dump big loads of snow that can demolish your campsite and collapse your tent. If you do camp beneath a tree, shake off the snow or use a long pole to beat it off before setting up camp.

It is best to clear the snow from your tent site several feet beyond the tent and the cooking area. Tents and kitchens set up on bare ground do not tilt wildly with the wind. Depressions beneath the tent can be stamped level with snow. If you sleep on a closed-cell foam pad, you won't be bothered by melting that's taking place beneath you.

Before leaving on the trip, check the waterproof bottom of your tent to be certain it doesn't have pinprick holes through which snow melt will invariably seep. Sealing compounds and patches are available at nearly all outdoor equipment stores. Check with the manufacturer's specifications for the best sealer. You can inspect the tent floor by holding it up to a bright light

Special equipment includes snow scoops to clear the campsite and tent stakes specially designed for anchoring lines in snow. Anodized-aluminum poles are almost essential for your tent so the snow and ice won't stick to them

For information on sleeping bags, see CHAPTER 5. You will need to extend the comfort range downward by either buying a heavier bag or adapting your summer-weight bag for winter use. This can be done by wearing heavy-duty underwear, such as high-quality wool or goose down. Your bag should be equipped with an insulated hood. You can also sleep with a wool cap or, in extreme temperatures, a down face mask that covers your entire head except for eye holes and slits for your nose and mouth.

Don't sleep in the buff. The nylon of the bag will be incredibly cold against your unprotected skin.

Zipping two bags together in cold weather is also an invitation to cold sleeping. It is almost impossible to keep cold air from seeping in around the shoulders.

The layering system of dressing is extremely important in winter camping, and careful shopping for clothing can give you a wide selection of warm clothing that individually isn't very heavy. A good camping wardrobe would include:

- net or wool underwear
- wool shirt
- wool sweater
- wool pants
- jacket or parka
- Dacron, nylon, or silk undersocks
- heavy wool outer socks
- waterproof boots with felt liners
- gaiters
- lightweight, water-repellent wind parka
- wind pants to match
- mittens with wool liners
- leather gloves with liners
- wool cap, balaclava style
- sunglasses

Individually, each of these items offers some protection from the cold and wet, and when you have them all on (with mittens and liners instead of gloves), you can comfortably survive temperatures well below freezing.

Your food should lean heavily toward carbohydrates and fat, to help your body produce heat, and lots of liquids.

Winter camping presents special hazards not covered in the chapter on first aid and safety, including thin ice and wet feet.

Ice: Many people fall through thin ice every winter. Avoid ice whenever possible, and search for shallow parts of streams to make your crossing. Learn to "read" ice and where to expect thin ice. It is usually thickest where streams are broadest and the water slower. On lakes it is thicker along the shore than farther out. If it is blue in color, it is usually thin.

Wet feet: If your feet get soaked in cold weather, stop immediately to build a fire and change socks. Frostbite and freezing occur rapidly, and the faster you get your feet dry and warm again, the less the danger.

Writer and guide Andy Russell tells

CAMPER'S SECRET

Everyone likes a bargain, and free campgrounds are one of the best bargains of them all. One couple who travels several weeks each year in their motorhome found a clever way to camp free, and often were the only campers in the park. They selected small towns along rivers where fishing prospects were good, and then talked to the local police. Frequently the police were happy to have someone in the town's park all night to help prevent vandalism. The couple had a CB radio in their motorhome, which they could use to summon the police if necessary. In several summers of free camping, they never had to use it.

of falling through the ice into a stream in the middle of an Alberta winter. He built a fire in record time and stripped off his woolen clothing that was already freezing stiff. When his companion arrived, his first sight was Russell standing stark naked in a snowbank, flailing his woolen long johns against a tree. With some justification, Russell's companion thought for a moment he was traveling with a demented woodsman. But Russell, having beaten the ice out of his woolen clothing, put it back on, and warmed himself by the fire, suffered no ill effects.

Obviously you should never be more than shouting distance from camp without carrying the Ten Essentials (see CHAPTER 3) and some kind of emergency shelter with you. In some parts of the country, blizzards can blow up almost instantly, and you may have to stop and hole up wherever you are to sit it out. If you have shelter and warm clothing and can build a fire and brew yourself a warm drink, you can sit out a storm in comfort.

The primary rule for winter camping is always to be prepared for the worst. Learn to live with winter instead of trying to force your schedule or your energies against it. You'll lose more such battles than you'll win. If the weather is foul, stay in camp. If bad weather is predicted, stay home. The more you understand winter camping, and your own limitations as well as your equipment's, the more you will enjoy this form of camping.

CHAPTER 12

WHERE TO NEXT?

One pitfall of instructional books on the outdoors is that the reader will become so overburdened with information that the purpose of the book—enjoyment of the outdoors—will be forgotten amid the rules, suggestions, charts, choices, and author's preferences (or prejudices). As more than one bewildered novice camper has said, books tell you how to do everything except have fun; they get you all dressed up with no place to go.

Thus we have departed slightly from the standard how-to approach in this chapter to give you an idea of what awaits the properly outfitted and educated camper. While most of the information in this book has been written on the assumption that the reader knows no more about camping than the average American knows about tribal customs in the Borneo highlands, by now you should feel a bit more comfortable in the maze of knowledge and equipment that technology and regulations have forced upon us.

Most campers who go to established campgrounds and camp within sight of their car soon get the urge to move on into the wilderness; to buy a pack and a pair of hiking boots, stock up on freeze-dried food and strike out through the forest or across the desert. Or they want to tackle a wilderness river, with either a guide service or a group of experienced whitewater people. The innate curiosity that has sent man all over the world and space soon becomes an urge you will want to follow.

This is the primary reason that

throughout the book we have continually nudged you toward purchasing lightweight equipment so that you won't have to duplicate tents, stoves, or sleeping bags when you want to try a different kind of trip. Buying the best and lightest equipment gives you a freedom of movement and choice you don't have if you buy according to price or convenience alone.

The basic camping equipment is universal in that you can use the same equipment in all seasons, all terrains, and all outdoor sports. Simply take the basic equipment and add the seasonal or terrain requirements to it. Winter camping requires additional clothing and warmer sleeping bags, for example, but the layer system of dressing takes care of most of the clothing, and you can increase the comfort range of your sleeping bag by adding a liner or one of those half-bags called bivouac bags for your lower body and a goose down-filled parka for your upper body.

This same basic outfit is all you will need for long river or lake trips, with the addition of waterproof duffel bags and life jackets.

And it can be used with equal success traveling in recreational vehicles. You may want to add a collection of books, maps, and food items such as corn-on-the-cob and any other kind of food you prefer, but you'll be glad you own sleeping bags that compress well: you can simply fold them into the stuff bag every morning.

After a few outings with your new equipment and your new form of recreation, you will quickly become accustomed to the outdoors, and if you have children they will soon be insisting that you take them on some real adventures. They will want something to tell their friends when school starts, something beyond the car camping that "everyone" does. They will want to go backpacking, or on a two-week bicycle trip. They will want to run a whitewater river, or take a long canoe trip that may involve portaging from lake to lake. Once you have started this form of recreation, the natural progression is for each trip to be a bit more rugged—and interesting—than the last.

All of us have our dream trips, those

CAMPER'S SECRET

Some of the best places for day trips and one-night camping trips are federal and state game refuges—but not during hunting season, of course. These refuges are usually near a source of good water, and camping areas are not difficult to find. Check with your state department of game or federal Fish and Wildlife Service for maps, because these areas are frequently overlooked by other campers.

journeys we would love to take if we had the time, and as soon as possible and before the children leave home for good.

Here are some examples of trips taken by one family that can be duplicated or applied to other parts of North America. These were taken in the American West and northern Canada and the Yukon Territory where some of the most beautiful wilderness in the world can be found. And all of these trips were taken with the basic equipment found in this book.

HIKING

Most of the Olympic Coast of Washington is enclosed in the coastal strip of Olympic National Park and is not touched by highway, logging road, commercial development, or campground. It is wild, subject to torrential rainfall and overcast days as well as beautiful sunny weather that exhibits the offshore rocks, the rugged headlands, and many natural arches carved out of the shoreline by wind and sea. The series of headlands can be walked around only at low tide; otherwise hikers must scramble over them on steep trails that often require dangling ropes for safe climbing.

The best time to make the trip is during a series of extremely low tides so you will have an opportunity to walk around the tide pools to examine the starfish, sea anemones, sea urchins, hermit crabs, barnacles, the hundreds of varieties of seaweed, and the brilliant-colored algae that stain the rocks several shades of green.

The National Park Service has been insistent on keeping the coast wild and has not established campsites anywhere along the way. Obviously, the most popular campsites are near the small streams that trickle down from the Olympic Mountains. Much of the water is stained brown from the decaying wood and other plant life it flows across. It must be boiled for purity, then left to stand a few more minutes so the sediment will collect on the bottom of the pot or pan.

Hikes along the coast can last as long as you want, from a single day to a week or longer, and it is possible to walk up to fifty miles along the beach in one direction with only one brush with civilization along the route. The hike is a combination of long nights with the sound of the surf against the rocks and on the sand and gravel beaches, and occasional heavy surf when a storm blows in off the Pacific.

One point of interest is the former archaeological dig on the northern end of the hike at Cape Alava. Here the Makah Indians lived for centuries in a village protected from the storms by a long shelf of a beach and a small, tree-topped island just offshore. The Makahs were famous whalers, and the archaeologists found a treasure trove of artwork in the village. The village was subject to infrequent but disastrous landslides from the cliff above, and each slide covered several of the cedar houses, burying the houses and occupants and keeping the artifacts intact beneath the solid covering of clay and soil.

Many hikers along the beach trail do not wear standard hiking boots. Instead, they prefer rubber boots for wading through the surf when it is necessary, and walking shoes or running shoes on the sand and gravel.

All the basic camping equipment discussed earlier is suitable for the trip, especially the lightweight equipment plus a standard backpack.

Another hike, which some refer to

194 FAMILY CAMPING

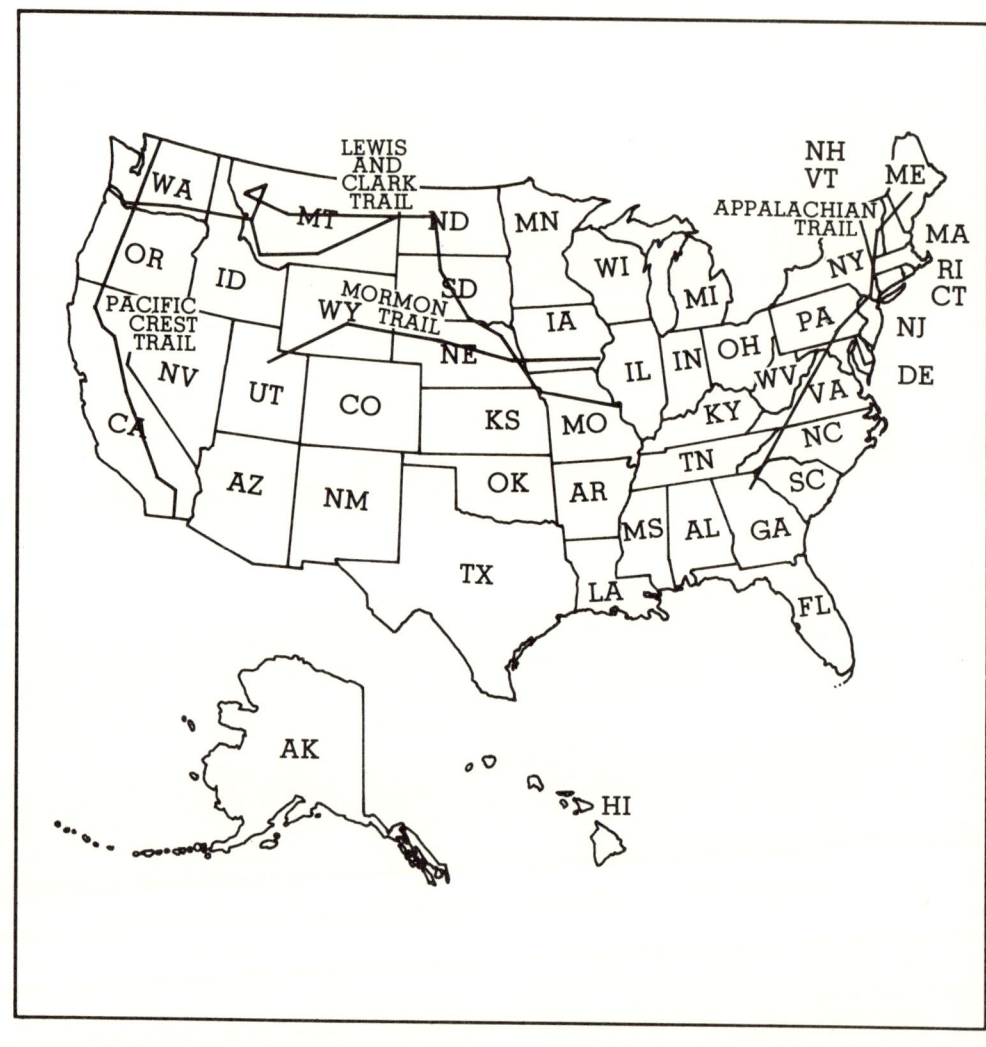

as a "classic" hike, is over the Chilkoot Trail from near Skagway, Alaska, to the headwaters of the Yukon River at Lake Bennett, British Columbia. The thirty-four-mile trail was the only overland portion of the route from the United States to the gold fields along the Klondike River near Dawson City, and during the gold rush of 1897 - 98, more than 20,000 persons walked the Chilkoot carrying their equipment from tidewater to the headwater lakes. The Northwest Mounted Police of Canada required each person to bring in a year's supply of food, clothing, and tools, and this translated to roughly a ton of gear per person.

The gold rush ended quickly, but during the two years that people—men, women, children, and assorted pets and work animals—traversed the trail, an amazing civilization rose and fell in that period. Six different towns sprang up almost overnight—Dyea at tidewater, Canyon City at the head of canoe navigation on the Tiaya River, Sheep Camp just below timberline, The Scales at the foot of the incredibily steep pass, Lindeman City at the lake of the same name, and Lake Bennett, where most of the Klondike stampeders spent the winter and spring building more than 7,000 boats and rafts while waiting for the ice to clear from the lakes and Yukon River so they could continue on north to the Klondike.

Today a hike over the trail is like walking through one of the world's longest museums. Bits and pieces of all the towns remain beside the trail, and one of the greatest of all engineering feats of the day is shown there, too. This is the aerial tramway that was built from Canyon City all the way over the summit to a small lake on the Canadian side, a total of ten miles. Some of the supports for the cable still stand, and sprockets, lengths of cable, and other parts lie beside the trail. At Canyon City are the remains of the steam power plant needed to operate the tramway.

Shoes, cookstoves, broken lanterns, collapsed cabins, lost or abandoned tools, a cache of canvas-and-wood collapsible canoes, harness, wagon parts and hundreds of other artifacts are still lying where they were dropped during the stampede.

The hike takes most people about five days to make, not because it can't be done in less time but because they want time to stop and explore the ghost towns and the other remnants of the gold rush. Excellent camping sites are found on both sides of the summit, and at the end of the trail on Lake Bennett, hikers can catch the White Pass & Yukon Route noon passenger trains that stop at Bennett Station to serve lunch to passengers and train crews. There they can continue northward to Whitehorse, the Yukon's capital, or go back over White Pass to Skagway, where they began.

OTHER HIKES

Two great Lower 48 trails: Each spring—sometimes more like late winter—several people carrying enormous loads in their packs strike out on an expedition that must be high on the dream list of every dedicated backpacker. These people are hiking one of the two great trail systems in America—the Appalachian and the Pacific Crest trails.

These are two of the premier trails in America, and fortunately they can be hiked in small, manageable portions instead of everyone having to take on the entire route.

The Appalachian Trail follows the crest of the Appalachian Mountains through 14 states. It begins to the south at Springer Mountain in northern Georgia and ends some 2,000 miles later at Mt. Katahdin in central Maine. It passes through Georgia, North Carolina, Tennessee, Virginia, West Virginia, Maryland, Pennsylvania, Vermont, New Hampshire, and Maine.

Since the Appalachians are not an unbroken range, the trail occasionally is forced to follow a highway short distances, which helps its hikers take breaks to repair shoes, purchase more food, and contemplate whether they want to continue or not.

Although it has been hiked in three months, perhaps less by some youthful charger, the average seems to be four months.

The Pacific Crest Trail is slightly longer, 2,400 miles, and traverses considerably more wilderness than the Appalachian. It goes through higher mountain ranges, across the Great American Desert, and hikes usually begin and end in winter.

The trail begins to the north in British Columbia's Manning Provincial Park, which adjoins North Cascades National Park across the international boundary in Washington. Glacier Peak Wilderness, skirts the edge of Mt. Rainier National Park, and crosses the Columbia River into Oregon at Bonneville Dam.

It continues along the ridges of the Cascades through Oregon into California, where the Cascades end and the Sierra Nevada Range begins. It swings southwest from the Sierra almost to Los Angeles, then swings east almost to the Joshua Tree National Monument, and stays in the desert until it reaches the Mexican border 30 miles east of San Diego.

Neither trail is easy to hike, and each has its own unique problems to conquer. Each has sections that are easy to hike; each has very tough stretches.

THE RIVERS

America has an abundance of great rivers for great float trips, ranging from the wilderness rivers of the West to the tamed giant rivers such as the Missouri, Mississippi, and Ohio and the smaller rivers with relatively short stretches of wilderness characteristics and great whitewater.

Some can be navigated by a novice in an inflatable raft or even a rented houseboat. But the most "interesting" rivers, meaning those with whitewater rapids and wilderness settings, should be attempted only by the very experienced or with a guide service that specializes in these rivers.

You can be as versatile on your river trips as you want. The large inflatable rafts that are so popular in many of the wilder rivers can carry enormous loads of people and gear, and you don't have to take freeze-dried food along unless you want to.

A memorable trip can be a gourmet float trip with the greatest concoctions you can dream up that are stored in the ice chest in one of the inflatables. You can carry tents large enough for standing, and you can take a folding table and folding lawn chairs if you want.

But in order to get the best "action" from the river it is best to take smaller inflatables that will crash through and over the rapids to give you a more thrilling ride. Some outfitters specialize in renting the small one- or two-man inflatable canoes with an experienced boatman as a leader to keep the incidence of dunkings to a minimum.

While traveling in these smaller craft, your lightweight camping gear will be especially appreciated because you can load all your belongings into a single waterproof duffel bag, and your

cameras, binoculars, and other similar items in one of the surplus military ammunition cans that are purchased by river outfitters by the hundreds.

Some of the best and longest rivers for whitewater trips are in the West, including Colorado, Wyoming, Montana, Utah, Idaho, Washington, Oregon, and California.

Smaller rivers with rapids are found in the South and Northeast, and a few are scattered through the Midwest. Some of the wildest, in terms of population centers, are in Alaska and western Canada.

One of the longest of the wilderness rivers is the Yukon, which begins in northern British Columbia on the Yukon border and winds its way north to Dawson city, Yukon, before turning west across Alaska, then southeast to its final run to the sea. The best portion for an interesting trip is from headwaters just south of Whitehorse, Yukon, to Dawson city, some 500 miles away. This is the last easily accessible place along the river for several more hundred miles. Rental canoes can be picked up at Dawson City, or you can have your own shipped back out more easily there than anywhere else along the river.

THE LAKE VOYAGES

Few manmade lakes are really suitable for wilderness-type trips because most are so thoroughly developed and heavily used. However, they are good places to become familiar with your equipment and for your family or group to become accustomed to water travel.

The best natural lakes—almost the only natural lakes in America—are in the northern states where the glaciers from the ice ages scoured out trenches that became lakes when the glaciers receded back into Canada and the Arctic.

Wisconsin, Minnesota, Michigan, Pennsylvania, New York, New Hampshire, Vermont, and Maine have the bulk of natural lakes that are not overcrowded with speedboats all through the summer. Each state has its own good selection of lakes for canoe trips. Occasionally you will find a group of lakes near enough together for you to portage from lake to lake, sometimes enabling you to make a complete circle in your trip.

These thousands of lakes continue on north into Canada, and the maps are dotted with them all the way from below the American border to the Arctic Ocean and Hudson's Bay.

One major canoe course is in the Voyagers National Park in northern Wisconsin along the Ontario, Canada, border. Strung all along that border are numerous lakes, many connected by shallow, slow streams that make it an ideal trip for canoeists.

RV TRIPS

Not all great trips require constant use of muscle power for transportation, and many families like to go on long trips in RVs when distance is the primary consideration.

One of the greatest trips for a family in an RV is to follow the Lewis and Clark Trail from St. Louis to the Pacific Ocean near Astoria, Oregon. The purists can point out that since Lewis began his trip in Pittsburgh, Pennsylvania, and met Clark along the Ohio River, that the trip really began on the East Coast. Thus, to follow the entire route, you would start at Pittsburgh, follow the Ohio River to the Mississippi, then go north along it to St. Louis (where historians generally say the expedition really began).

From St. Louis, the party followed the Missouri River into southwestern Montana, then across Idaho, down the Snake River between Washington and Oregon, and at last to the Pacific.

The Lewis and Clark Trail is perhaps the best route for seeing the northern portion of the Great Plains, the Rocky Mountains, and the Pacific Northwest. The Missouri River flows through Missouri, Kansas, Iowa, Nebraska, South Dakota, North Dakota, and Montana.

From there, the explorers crossed a low pass into central Idaho near Salmon, went on north again back into Montana, and crossed back into Idaho over Lolo Pass and followed the Clearwater River down the Snake River at Lewiston, Idaho. Then the route goes through the southeast corner of Washington, and between the two states along the Columbia River to the sea.

One of the bonuses of this route is that when you reach Montana, you aren't far from two of the most popular parks in America—Yellowstone to the south and Glacier to the north.

All along the route are other sites of historical and scenic interest, such as old army posts, trading posts, Indian reservations, and some of America's most beautiful scenery.

RVs are also an excellent way to travel through some of the more beautiful scenery that does not lend itself to camping. This is true of much of America's coastline which has ample commercial campgrounds, but few (and quite crowded) places suitable for pitching tents. Most of these can be reached only by hiking some distance, and you have to plan your trip quite carefully and take risks of finding available campsites. Also, you must take a chance and leave your car behind at the trailheads.

Thus, camping in these areas in RVs is much simpler and safer.

Again, all the basic equipment you have bought for the family camping trips is suitable for RVs, even those which have microwave ovens, televisions, food processors, and all the other modern conveniences. The children can get away from the adults (or vice versa) by pitching a tent just outside the RV door. The lightweight sleeping bags take up much less room than the heavier so-called slumber bags, and you'll invariably find good uses for the other equipment you bought for car camping.

In the following section are listings of all the major public agencies involved in camping. Also, you will find some basic information on different regions of America and other things of interest to campers.

Enjoy.

PART IV
RESOURCES

CHAPTER 13

DESTINATIONS

PRIVATE OR GOVERNMENT CAMPGROUNDS

Most families develop their own list of favorite camping sites over a period of time, some of which may be nothing more exotic than a level spot on a bank overlooking a stream. Yet, owing to the demand for such private sites, you will likely spend a portion of your time in either privately owned or government campgrounds.

Some privately owned campgrounds are part of nationwide chains, each site operated by a franchise owner. Most of these chains have well-defined guidelines for the franchise owner so that you know what to expect each time you pull into the office and register for a stay. These welcome everyone from car campers with their own tents to motor-home campers.

They will have a combination office, grocery store, and game room. They will also have shower rooms and a coin-operated laundry. Campsites for self-contained RVs will have an electrical outlet, a water faucet, and a sewage hookup. You must carry your own electrical cord—and always carry a variety of plug adapters, because there is no standard plug; some require two-pronged plugs, others three-pronged ones, and some oversized plugs similar to those on an automatic dryer or range.

You must also carry your own section of water hose and sewage hose. If you don't want to be hooked to the sewage system, you can use the RV's holding tank until you prepare to leave the following morning. Nearly all privately owned campgrounds have a special place set aside for dumping sewage, usually near the entry area.

These campgrounds are usually near interstate and major highway interchanges, and an increasing number of car campers use them for their convenience. Renting a tent site is considerably cheaper than an RV site with the electrical, water, and sewage hookups, and it is an inexpensive way to travel across country, much cheaper than staying in motels and eating all meals in restaurants.

A few campgrounds have tents for rent, but these are very few in number, and you should take no chances. Always carry your own tent.

Government campgrounds are usually less expensive than privately owned ones, but they are also usually more primitive, less convenient to find, and farther away from population centers. The larger ones, however, sometimes have a small store and service station nearby.

These campsites range from small county parks to those developed by various state agencies, and federal government parks developed by the Bureau of Land Management, the Forest Service, and the National Park Service.

Canada has a wide selection of excellent parks in the national system, Parks Canada, and each province has its own selection of parks.

At the most primitive end of the scale, you will find a leveled area with firepits or concrete blocks with steel grates, a few garbage cans, and an outhouse. A water supply is usually nearby, either a stream or lake, or a well with a hand pump. Unlike the larger or better-developed parks, these frequently have no time limit on your stay and are often free.

Most states have a good selection of state parks with a resident ranger and well-developed campsites that include water and electrical hookups. These will also have recreation areas where you can play volleyball and badminton, pitch horseshoes, or enjoy other forms of entertainment. Nearly all will have picnic tables at each site. Usually you must make reservations well in advance for campsites, although a few continue to operate on a first-come, first-served basis. During the peak summer season, it is obviously best to arrive at these first-come, first-served campgrounds just before checkout time so you can claim a campsite, then walk back to the ranger's office to pay.

The same policy prevails at most Forest Service campgrounds, and many have a self-payment system.

The Forest Service has established a rating system on a 1-to-5 scale, with the most primitive campgrounds rated 1. This system is as follows.

1. Very primitive: These are primarily trail camps in wilderness and primitive areas, and usually are reached only on foot or horseback. Although some have stone or concrete firepits, usually they are little more than clearings on high ground. They will be shown on the maps you should purchase from the ranger station or district headquarters.

2. Rugged: Only a bit more developed than primitive sites, these can usually be reached by rough roads in vehicles that are not towing trailers. No established campsites are marked, and usually no outhouses or other such structures are present.

3. Rustic: These usually appear primitive but have a few amenities, such as outhouses. Campsites are usually laid out so that you'll have only two or three per acre, giving you plenty of room for privacy. Trails and roads into them are usually maintained well enough for towing a trailer to them and are either all-weather gravel or dirt.

4. Modern: Hard-surfaced roads lead into these, and you'll find good latrines and fire facilities. Usually these have three to five campsites per acre and often have some kind of natural screening between sites. These campgrounds are often near a recreational area, such as a stream or lake. Nature trails are frequently established nearby.

5. Ultramodern: These are the top of the Forest Service line, with laundry facilities, community showers and bathhouses, flush toilets, electrical and water hookups, landscaped grounds, and a ranger assigned to the campground.

Most Forest Service campgrounds are free, except the "ultramodern" ones, which charge a nominal fee.

You are advised to obtain information on campgrounds in all national forests you plan to visit, and do so well in advance, since many operate only on the reservation system. A list of the major national forests throughout the United States is in the next section.

Most major national parks have well-maintained and well-patrolled campgrounds, usually with a daily schedule of activities available to visitors. These range from organized nature hikes with a park naturalist to evening slide shows and lectures on the park. Usually the national park campgrounds are on a par with the best of the privately owned campgrounds. Privacy is often a low priority because the parks are so heavily used that the park planners must crowd as many campsites into a limited area as possible. In exchange for your fee, you receive some security, the use of the bathhouse and flush toilets, and the convenience of being near the attractions of the park itself. Nearly all of these operate on a reservation system, and your stay will be limited to a specific number of days.

ADDRESSES

After you have gone camping a few times and made some long trips to sample a variety of campgrounds, you will soon develop a sixth sense to detect areas where you will want to camp in the future, as well as places you will want to avoid. You will learn which public land agencies have the best facilities for your requirements, and which national forests have a reputation for taking good care of their campgrounds.

Soon you will find yourself going to the same general areas again and again, in part because you know what to expect from previous visits, and also because we are creatures of habit and like familiar surroundings.

Following is a list of addresses for information on national parks, national forests, and state camping areas in the United States. National park maps are free; national forest maps, which are excellent and beautiful enough to frame for the recreation room, cost $1 each. Most state agencies have free maps that range from full-color detailed maps to rough reading of main roads.

The best sources for information on these camping areas are three books published by Rand McNally & Co.:

Campground and Trailer Park Guide, revised annually.

National Park Guide, by Michael Frome, also revised annually.

National Forest Guide, by Len Hilts, revised as necessary.

Outdoor equipment stores and bookstores in each area will stock regionally published books, but the three above will serve you well.

Although there are several national campground organizations, one has outrun all the others in establishing a chain of franchised campgrounds with uniform standards throughout all of North America, from the Yukon to Mexico. This is Kampgrounds of America (KOA). Its directory of campgrounds is free at any KOA franchise, or available for $1 from KOA, Inc., Billings, MT 59224.

For information on other privately owned and operated campgrounds, contact the National Campground Owners Association, 804 D Street N.E., Washington, DC 20002.

One of the best sources of information on all campgrounds is the American Automobile Association; your membership in any state association entitles you to the detailed guides of any state or area of North America.

Still another good campground guide is that published annually by Woodall Publishing Co.

The best source for information on camping in Alaska, British Columbia, and the Yukon Territory is the annual **Milepost,** published annually by Alaska Northwest Publishing Co.

Many other campgrounds can be found by contacting state and federal fish, game, and land management agencies in the areas you plan to visit.

Several large timber companies also offer free camping and provide free maps showing locations. Some major utility companies with holdings along artificial lakes for hydroelectric projects also have free campgrounds.

EASTERN NATIONAL PARKS

North Atlantic Region
National Park Service
15 State St.
Boston, MA 02109

Mid-Atlantic Region
National Park Service
143 South Third St.
Philadelphia, PA 19106

National Capital Region
National Park Service
1100 Ohio Drive, S.W.
Washington, DC 20242

Southeast Region
National Park Service
1895 Phoenix Blvd.
Atlanta, GA 30349

MIDWESTERN PARKS

Midwest Region
National Park Service
1709 Jackson St.
Omaha, NE 68102

WESTERN PARKS

Western Region
National Park Service
450 Golden Gate Ave.
San Francisco, CA 94102

Southwest Region
National Park Service
Old Santa Fe Trail
P.O. Box 728
Santa Fe, NM 87501

Pacific Northwest Region
National Park Service
Westin Bldg.
Seattle, WA 98121

Rocky Mountain Region
National Park Service
655 Parfet St.
P.O. Box 25287
Lakewood, CO 80225

NATIONAL FORESTS

Eastern Region
National Forest Service
633 West Wisconsin Ave.
Milwaukee, WI 53203

Southern Region
National Forest Service
1720 Peachtree Rd., S.E.
Atlanta, GA 30309

Alaska Region
National Forest Service
RR 4, Box 1628
Juneau, AK 99801

California Region
National Forest Service
630 Sansome St.
San Francisco, CA 94111

Intermountain Region
National Forest Service
324 25th St.
Ogden, UT 84401

Northern Region
National Forest Service
Federal Bldg.
Missoula, MT 59801

Pacific Northwest Region
National Forest Service
319 S.W. Pine St.
P.O. Box 3623
Portland, OR 97208

Rocky Mountain Region
National Forest Service
11177 West 8th Ave.
Box 25127
Lakewood, CO 80225

Southwestern Region
National Forest Service
517 Gold Ave. S.W.
Albuquerque, NM 87102

EASTERN AND SOUTHEASTERN STATE CAMPING AREAS

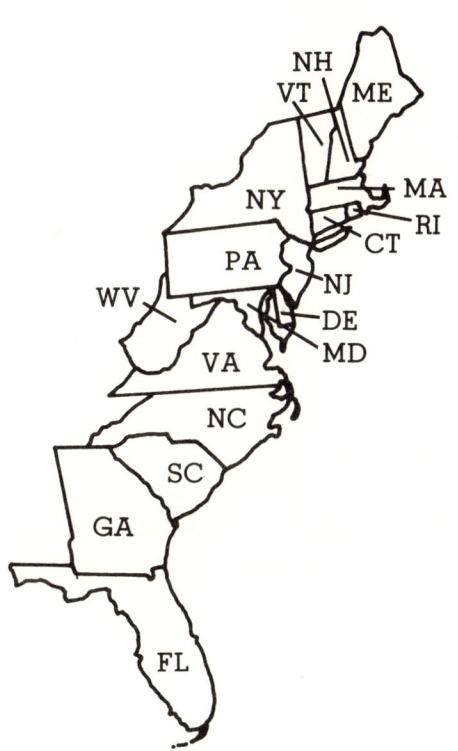

MAINE
Bureau of Parks & Recreation
State Office Bldg.
Augusta, ME 04333

NEW HAMPSHIRE
Office of Vacation Travel
P.O. Box 856
Concord, NH 03301

VERMONT
Dept. of Forests, Parks and Recreation
Montpelier, VT 05602

MASSACHUSETTS
Dept. of Environmental Management
Div. of Forests and Parks
100 Cambridge St.
Boston, MA 02202

CONNECTICUT
Dept. of Environmental Protection
Office of Parks and Recreation
165 Capitol Ave.
Hartford, CT 06115

NEW YORK
Bureau of Communications
N.Y. State Office
of Parks and Recreation
Empire State Plaza
Albany, NY 12238

NEW JERSEY
Dept. of Environmental Protection
Div. of Parks and Forestry
CN404
Trenton, NJ 08625

PENNSYLVANIA
Bureau of State Parks
Dept. of Environmental Resources
P.O. Box 1467
Harrisburg, PA 17120

RHODE ISLAND
Div. of Parks and Recreation
83 Park St.
Providence, RI 02903

MARYLAND
Maryland Park Service
Tawes State Office Bldg.
Annapolis, MD 21401

DELAWARE
Dept. of Natural Resources
and Environmental Control
Div. of Parks & Recreation
P.O. Box 1401
Dover, DE 19901

VIRGINIA
Div. of State Parks
1201 Washington Bldg.
Richmond, VA 23219

WEST VIRGINIA
Parks and Recreation Division
Dept. of Natural Resources
State Capitol Bldg. No. 3, Rm. 311
Charleston, WV 25305

NORTH CAROLINA
Div. of Parks and Recreation
Dept. Natural Resources and
Community Development
P.O. Box 27687
Raleigh, NC 27611

SOUTH CAROLINA
Div. of Parks and Recreation
and Tourism
Rm. 30, Box 71
Columbia, SC 29202

FLORIDA
Florida Dept. of Natural Resources
Bureau of Education Information
3900 Commonwealth Blvd.
Tallahassee, FL 23200

GEORGIA
Office of Information
Dept. of Natural Resources
270 Washington St., Rm. 817
Atlanta, GA 30334

SOUTHERN STATE CAMPING AREAS

ALABAMA
Alabama Dept. of Conservation
Div. of State Parks
Administrative Bldg.
Montgomery, AL 36130

ARKANSAS
Dept. of Parks and Tourism
One Capitol Mall
Little Rock, AR 72201

LOUISIANA
Louisiana Office of State Parks
P.O. Drawer 1111
Baton Rouge, LA 70821

MISSISSIPPI
Travel and Tourism Dept.
Mississippi Dept.
of Economic Development
P.O. Box 849
Jackson, MS 39205

TENNESSEE
Dept. of Conservation
Div. of State Parks
2611 West End Ave.
Nashville, TN 37203

KENTUCKY
Travel
Frankfort, KY 40601

MIDWESTERN STATE CAMPING AREAS

MISSOURI
Missouri Dept. of Natural Resources
P.O. Box 176
Jefferson City, MO 65102

OHIO
Publications Center
Ohio Dept. of Natural Resources
Fountain Square Bldg. B
Columbus, OH 43224

ILLINOIS
Illinois Adventure Center
160 North LaSalle St.
Chicago, IL 60601

INDIANA
Dept. of Natural Resources
Div. of State Parks
616 State Office Bldg.
Indianapolis, IN 46204

MICHIGAN
Parks Div.
Dept. of Natural Resources
P.O. Box 30028
Lansing, MI 48909

MINNESOTA
Div. of Parks and Recreation
Box 39, Centennial Bldg.
St. Paul, MN 55155

WISCONSIN
Dept. of Natural Resources
Bureau of Parks and Recreation
Box 7921
Madison, WI 53707

IOWA
State Conservation Commission
Henry A. Wallace Bldg.
Des Moines, IA 50319

NEBRASKA
Game and Parks Commission
22000 North 33rd St.
P.O. Box 30370
Lincoln, NE 68503

SOUTH DAKOTA
Game, Fish and Parks Dept.
Div. of Parks and Recreation
Anderson Bldg.
Pierre, SD 57501

NORTH DAKOTA
North Dakota Park Service
Pinehurst Office Bldg.
1424 W. Century Ave.
Bismarck, ND 58502

GREAT PLAINS STATES CAMPING AREAS

SOUTHWESTERN STATES CAMPING AREAS

OKLAHOMA
Oklahoma Tourism and Recreation Dept.
Div. of Marketing Services
500 Will Rogers Bldg.
Oklahoma City, OK 73105

KANSAS
Parks and Resources Authority
503 Kansas Ave.
P.O. Box 977
Topeka, KS 66601

TEXAS
Texas Parks and Wildlife Dept.
42000 Smith School Rd.
Austin, TX 78744

NEW MEXICO
State Park and Recreation Div.
Box 1147
Santa Fe, NM 87503

ARIZONA
Arizona Office of Tourism
3507 N. Central, Suite 506
Phoenix, AZ 85004

NEVADA
Dept. of Conservation and
Natural Resources
Nevada Div. of State Parks
Capitol Complex
Carson City, NV 89710

ROCKY MOUNTAIN STATES CAMPING AREAS

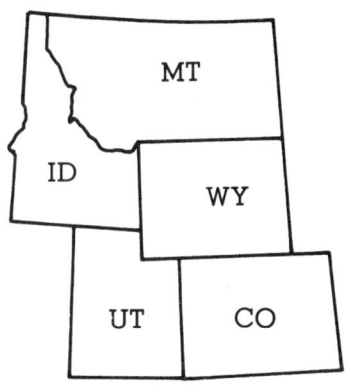

COLORADO
Div. of Parks and Recreation
618 Centennial Bldg.
1313 Sherman St.
Denver, CO 80203

WYOMING
Wyoming Recreation Commission
604 E. 25th St.
Cheyenne, WY 82002

MONTANA
Dept. of Fish, Wildlife and Parks
Parks Div.
1420 East Sixth Ave.
Helena, MT 56901

IDAHO
Parks and Recreation Dept.
Statehouse Mall
2177 Warm Springs
Boise, ID 83720

UTAH
Parks and Recreation
1596 West North Temple
Salt Lake City, UT 84116

PACIFIC COAST STATES CAMPING AREAS

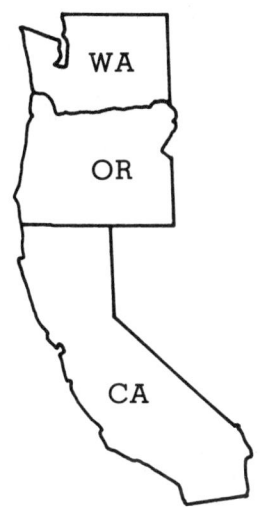

CALIFORNIA
Dept. of Parks and Recreation
P.O. Box 2390
Sacramento, CA 95811

OREGON
Travel Information
Oregon Dept. of Transportation
101 Transportation Bldg.
Salem, OR 97310

WASHINGTON
State Parks and Recreation Commission
7150 Clearwater Lane
Olympia, WA 98504

212 FAMILY CAMPING

ALASKA
Div. of Parks
619 Warehouse, Suite 210
Anchorage, AK 99510

HAWAII
Dept. of Land and Natural Resources
State Parks Div.
1151 Punchbowl St., Rm. 310
Honolulu, HI 96813

CANADIAN NATIONAL PARKS

Parks Canada
Dept. of Indian Affairs
Centennial Tower
400 Laurier Ave. West
Ottawa, Ontario, Canada K1R 5C6

PROVINCIAL PARKS

EASTERN CANADA

Dept. of Tourism
Promotion Branch
P.O. Box 12345
Fredericton, New Brunswick
E3B 5C3

NEWFOUNDLAND
Dept. of Tourism
130 Water St.
St. John's, Newfoundland
A1C 1A8

ONTARIO
Ontario Travel
Queen's Park
Toronto, Ontario
M7A 2E5

PRINCE EDWARD ISLAND
Visitor Services Division
P.O. Box 940
Charlottetown, Prince Edward Is.
C1A 7M5

QUEBEC
Parks Quebec
150 St. Cyrille Blvd., E, 10th Floor
Quebec, Quebec
G1R 4Y1

WESTERN CANADA

ALBERTA
Travel Alberta
Alberta Tourixm and Small Business
Box 2500
Edmonton, Alberta
T5J 2Z4

BRITISH COLUMBIA
Ministry of Tourism
1117 Wharf Street
Victoria, British Columbia
V8W 2Z2

MANITOBA
Dept. of Economic
Development and Tourism
Travel Manitoba
Dept. 2044, Legislative Bldg.
Winnipeg, Manitoba
R3C OV8

NORTHWEST TERRITORIES
Travel Arctic
Yellowknife
Northwest Territories
X1A 2L9

SASKATCHEWAN
SaskTravel
Dept. of Tourism and
Renewable Resources
3211 Albert St.
Regina, Saskatchewan
S4S 5W6

YUKON
Parks and Historic Resources
Box 2703
Whitehorse, Yukon
Y1A 2C6

MAPS
For recreational maps of national parks and national forest service campgrounds, write or visit the offices nearest you, or write them at the addresses shown in the previous section.

For Geological Survey maps and the index, write any of the following offices, whichever is nearest you:

EAST OF THE MISSISSIPPI
U.S. Geological Survey
Map Distribution Center
1200 Eads St.
Arlington, VA 22202

WEST OF THE MISSISSIPPI
U.S. Geological Survey
Map Distribution Center
Federal Center Bldg.
Denver, CO 80225

CANADA
Canadian Map Office
Dept. of Energy, Mines and Resources
614 Booth St.
Ottawa, Ontario K1A 0E9

CHAPTER 14

FURTHER READING

Since the arrival of the first white men in North America, outdoor-adventure reading material has reflected Americans' great love of the outdoors; after all, we have more of it left intact than most other nations. Our national heritage is laden with great outdoor adventures that include journals of explorations toward the West, great sea voyages, reports by fur-trapping expeditions, first descents of rivers, and first ascents of mountains.

In addition to the great adventures, we have also had a number of great naturalists 'who described the world around us in terms we never tire of reading.

Since World War II Americans have been more and more interested in outdoor recreation, and a flood of how-to books, guidebooks, and other basic publications have come onto the market, many of which have remained in print for several generations of readers. All of this activity has made American campers among the most literate and well informed in the world.

Thus, the following bibliography is a mix-and-match of the outdoor adventure and the more straightforward instructional books.

Angier, Bradford. **Field Guide to Edible Wild Plants.** Stackpole, 1976. A heavily illustrated guide to wild plants, where to find them, and how to prepare them.

How to Stay Alive in the Woods. Collier, 1976. A basic book on survival that covers how to find food, shelter, and water and how to use both compass and celestial navigation.

Ferber, Peggy, ed. **Mountaineering: The Freedom of the Hills.** The Mountaineers, 1974. How-to for the advanced outdoorsman who will be doing some climbing as part of the outdoor experience.

Fletcher, Colin. **The New Complete Walker.** Alfred A. Knopf, 1974. An informative, often hilariously funny instructional book on backpacking.

From Katahdin to Springer Mountain. Rodale Press, 1977. A collection of stories of hikers along the Appalachian Trail, including grandmothers and young children.

Gibbon, Euell. **Stalking the Blue-Eyed Scallop.** David McKay, 1964.

Stalking the Wild Asparagus. David McKay, 1962.

Both books are standards by the American who single-handedly introduced wild foods to the nation.

Hillcourt, William. **The Official Boy Scout Handbook.** Boy Scouts of America, 1979. One of the best sources of outdoor how-to in existence.

Kjellstrom, Bjorn. **Be Expert with Map and Compass.** Charles Scribner's Sons, 1976. A basic guide to using compass and map.

Lathrop, Theodore G., M.D. **Hypothermia: Killer of the Unprepared.** The Mazamas, 1972. Basic information on hypothermia, how to avoid it, and how to treat it.

Leopold, Aldo. **A Sand Country Almanac.** Oxford University Press, 1979. One of the classics on enjoying the outdoors with an emphasis on environmental protection.

Manning, Harvey. **Backpacking: One Step at a Time.** Vintage Books, 1980. A guide to outfitting yourself for the outdoors with an emphasis on backpacking.

Olsen, Larry Dean. **Outdoor Survival Skills.** Brigham Young University Press, 1973. Food, shelter, water, and other information on surviving.

Rugge, John, and Davidson, James West. **The Complete Wilderness Paddler.** Alfred A. Knopf, 1977. A how-to book with excellent illustrations of how to outfit yourself for running whitewater rivers.

Rutstrum, Calvin. **The Wilderness Routefinder.** Collier Books, 1967. A book written with the wilderness traveler in mind, including celestial navigation and lots of personal observations.

Sutton, Ann and Myron. **The Pacific Crest Trail.** Lippincott, 1975. Not a guide to the trail (several of those exist), but a naturalist's description of the trail and the flora and fauna to be expected on the 2,400-mile trek.

Van Lear, Denise, ed. **The Best about Backpacking.** Sierra Club, 1974. Each of the chapters is by an expert on various subjects related to outdoor recreation.

Wilkerson, James A., M.D. **Medicine for Mountaineering.** The Mountaineers, 1979. Excellent information on all forms of outdoor survival and how to treat common ailments and injuries.

RESOURCES

One of the greatest adventures in American history was the Lewis and Clark Expedition of 1804-6, mentioned earlier as an excellent car camping or RV route across the West. You can consult works on the adventure, beginning with the eight-volume set of the journals kept by Capt. Meriwether Lewis and Lt. William Clark. The set is available from Arno Press and from antiquarian book dealers. Two other works of value are:

Cutright, Paul Russell. **Lewis and Clark, Pioneering Naturalists.** University of Illinois Press, 1969. Easily the most readable account of the expedition, this is written without the usual political-economic overtones of most studies.

Satterfield, Archie. **The Lewis and Clark Trail.** Stackpole, 1978. A combination history-guide to the expedition and trail today.

In addition, see the books listed in the Acknowledgments for further reading suggestions on a variety of outdoor subjects.

INDEX

A

Activities, camping
  for children, 10, 33, 143, 145–146, 154, 159, 165
  knots, 154–157
  nature scavenger hunt, 154
  orienteering, 152–153
  for rainy days, 146
  species identification, 143–144, 157
  stargazing, 144, 146, 150–151
  using a compass, 145, 148–152
  using knives, axes, saws, 159–165
  using topographical maps, 145, 146–148
  wildcrafting, 165
Addresses, 205–213
Air mattress, 48, 80, 81
Air pumps, 81
American Automobile Association (AAA), 206
Ammunition cans, 197
Animals, identifying, 143–144, 157–158. *See also* Pests, outdoor
Ants, 174
Appalachian Trail, 195
Arrowheads, 141
Axes, 47, 159–164

B

Backpacking
  advantages of, 16
  clothing for, 16, 85, 193
  sleeping bags for, 48, 74
  tents for, 16, 58, 59, 60, 69
  trips, 193–195
Backpacks, 53–54
"Balloon cloth," 43, 48
Bear grass, 141
Bears, 173–174
Bedding, 48, 71–81
Bees and hornets, 174
Biking, 16, 74, 85

Birds, 113
Bleeding, 175–176
Boating, 48, 50, 69, 74, 192, 196–198
Books
  on camping, 215–217
  on camping areas, 205–206
  on nature, 143–144, 157
Boots, 85–88
Boy Scouts of America Outdoor Code, 33
Burns, 169–170, 175

C

Cagoule, 93
Campfire grid, 48
Campfires
  building, 101–107
  fire-starting items, 46, 49
  putting out, 104, 106–107
  restrictions, 102, 116, 122, 128, 157
  safety, 100–101, 102–103, 104–105, 111, 159
  site of, 100, 106
  wood for, 34, 102, 105, 157
Campgrounds
  activities, 33, 143–153
  choosing, 13, 97–99
  curfews, 35
  fees, 28, 204
  free, 188
  government, 97–98, 203–205
  information, 205–206
  manners at, 10, 32, 102
  modern, 205
  noise in, 32, 35
  pests, 31, 113, 171–177
  pets, 31–32, 50, 177
  primitive, 204
  private, 203–205
  reservations at, 205
  vandalism in, 35
Camping
  in American history, 5–18, 39

Camping (*cont.*)
   benefits of, 1–3
   expenses, 28
   for first-timers, 15–35, 49, 80
   in spring and fall, 183–186
   types of, 16–17
   in winter, 186–189, 192
Camping trailers, 22–24
Campsites
   characteristics of good, 99, 100, 185, 187
   dividing the labor, 109–113, 127
   garbage disposal, 108–109
   leaving, 102, 107
   locating latrines, 101, 107–108
   marking, 92
   RV, 203–204
   setting up, 100–101, 109
   weather and, 185, 187
Campstoves. *See* Stoves
Canada, 194–195, 204, 212–213
Candles, 46, 48, 49, 52–53
Canoeing, 74, 197–198
Car camping, 16–17, 30, 48, 120
   check list for, 50, 54–55
Car-top carrier, 53
Cats, 31–32
Cattail, 140
Charcoal, 128
Chickweed, 140
Chiggers, 174
Children
   age of, 2, 9, 10
   activities for, 10, 33, 143, 145–146, 154, 159, 165
   bringing a friend, 10
   bringing favorite items, 35, 50
   campfires and, 102–103
   and choosing a campground, 13, 33
   clothing for, 85
   fears of, 32
   and outdoor manners, 10, 34
   tasks for, 9, 11, 109
   tents and, 59
Chilkoot Trail, 194–195
Clothing, 83–93
   for children, 85
   extra, 45–46, 178
   layering, 84, 188
   materials, 42, 85
   for spring and fall, 183–184
   weight of, 39
   for winter, 83, 186, 188, 192
   see also specific names of clothing
Coats. *See* Jackets; Parkas; Raingear; Windbreakers

Compass
   learning to use, 46, 145, 148–149
   orienteering, 152–153
   types, 149
   using with maps, 47, 146, 149, 152
   watch, 153–154
Cooking, outdoor, 115–122
   check list, 55
   chores, 110, 111, 116, 127–130
   kitchen location, 100–101
   menus, 122, 130–139
   methods, 106, 118, 124–127, 140
   tablewear, 49, 55
   utensils, 39, 43, 49, 55, 122–127
   see also Campfires; Foods; Stoves
Coolers
   efficiency of, 128, 135, 141
   packing, 54
   uses for, 122–124, 168
Cord and wire, 49
Cots, 80–81
Cross-country skiing, 74, 186, 187
Cuts, 175–176

D
Dandelion, 140
Desert areas, 99
Dishwashing, 116, 127–130
Dock, 140
Dogs, 31, 177
Down, 6–7, 42, 85
   care of, 74–75, 79
   defined, 72
   methods of construction, 77–78
   salvaging, 74–75
   substitutes for, 71–72
Duffel bag, 54, 192, 196
Dutch ovens, 124–127, 140

E
Electricity
   generators, 26–27
   safety standards, 29–31
   wattage draws, 26–27, 28
Elevations, higher, 178
Emergencies
   survival skills, 49–50, 157
   weather changes, 99, 178, 189
   see also First aid
Emergency flares, 174
Equipment
   basic needs, 15, 44, 45–46, 49
   choosing, 15, 30, 192
   Eddie Bauer's list, 47–49
   group needs, 45
   history of, 5–6

individual needs, 44
investment in, 40–42
unnecessary, 45

F
Fall camping, 183–186
Fees, campground, 28, 204
Fifth-wheel trailers, 25–26
File, 49
Fire, RV safety standards, 29. *See also* Campfires
Fire-starting items, 46, 47, 49
Firewood, 34, 102, 105, 157
First aid
   bleeding, 175–176
   burns, 169–170, 175
   cuts, 176–177
   fractures, 175
   frostbite, 167
   heat exhaustion, 170–171
   hypothermia, 178–182
   outdoor pests, 171–175
   poisonous plants, 177–178
   shock, 174–175
   snakebite, 168–169
   stretchers, 174
   sunstroke, 170–171
First aid kit, 46, 48–49, 167, 168–171
Flashlight, 46, 48, 51–53, 113
   spare batteries/bulbs, 46, 51, 113
Flyrod, 47
Foods
   dehydrated and freeze-dried, 39, 45, 49, 110, 130, 139
   extra, 45, 49
   nutritious, 131
   snack items, 139, 185–186
   for spring and fall camping, 185–186
   storage at campsites, 100, 113, 173
   wild, 139–141, 177
   for winter camping, 188
Footwear, 42, 85–88, 184, 193
Forest products, 165
Forest Service, 204
Four-wheel-drive vehicles, 18–19
Fractures, 176
Frostbite, 167, 188
Fuel consumption, 17, 20, 21, 27–28

G
Game refuges, 192
Garbage, 108–109, 128–130
Gas mileage. *See* Fuel consumption
Generators, 26–27
Glacier National Park, 173, 198
Glasses, 46, 50

Gloves, 90, 184, 185
Golden Age Passport, 26
Golden Eagle Passport, 26
Grease, 109
Griddles, 124
Grills, 124
Grommets, ball-and-wire, 67, 68, 92
Gross vehicle weight (GVW), 21, 22, 27

H
Hammocks, 81
Hand warmers, 121–122
Headgear, 90–91, 93, 185, 188
Heaters, 58, 121–122
Heat exhaustion/stroke, 170–171
Hiking trips, 74, 193–196. *See also* Backpacking
Hitches, 24–28
Home, securing, 55
Hookups, 23, 203–204
Horse-packing trips, 58
Hypothermia, 83, 178–182

I
Ice, dangers of, 188–189
Insect repellent, 171, 174
Insects. *See* Pests, outdoor
Insect spray, 173, 174
Itching, 177–178

J
Jackets, 88–90

K
Kampgrounds of America (KOA), 206
Kitchen area, 100–101. *See also* Cooking, outdoor
Knives, 46, 47, 49, 160
   safety and, 159
   and sharpening tool, 46, 47
Knots, 154–157

L
Lake voyages, 197–198
Lamb's quarters, 140
Lanterns, 16, 35, 51–53
Latrines, 101, 107–108
Laundry, 88, 110
Lewis and Clark Trail, 198–199
Loft, 72–74
LP gas systems, 29

M
Manners, outdoor, 32, 33–35
   children and, 10, 34
   leaving a campsite, 102, 107

Manning Provincial Park, 195–196
Maps
    Geological Survey, 146, 149, 213
    purchasing, 146, 213
    road, 46
    topographical, 46, 47, 145, 146–148, 152, 213
    using a compass with, 149
Matches, 46, 47, 49
Medicines, 46
Menus, 122, 130–139
Milepost, 206
Mini-motorhomes, 20–21
Mosquitoes, 174
Motorcycle camping, 16
Motorhomes, 17–18, 20–21
Mountaineering, 74, 186
Mt. Rainier National Park, 196
Mushrooms, 139–140

N

Nails, 49
National Forests, 8, 33, 97, 207
National Parks, 98, 173, 193, 205, 206–207
National Park Service, 8, 31, 86, 98, 193, 204
Navigation, basics of, 145–146, 149
Nettles, 140–141, 177–178
Nightwear, 88
Noise, 32, 35
North Cascades National Park, 196
Nutrition, 131

O

Olympic National Park, 193
Orienteering, 152–153
Ovens, 124–127

P

Pacific Crest Trail, 195–196
Packboards, 47–48
Packhorse, 48
Packing, 53–54
Pads, 79–80, 187
Pants, 88
Parkas, 88–90
Personal items, 50
Pests, outdoor, 31
    ants, 174
    bears, 173–174
    birds, 113
    chiggers and noseeums, 174
    flies, 172
    and food, 113
    and latrine location, 108
    mosquitoes, 174
    porcupines, 174–175
    skunks, 174
    slugs and snails, 171
    spiders, 172–173
    stinging insects, 172, 174
    ticks, 171–172
Pets, 31–32, 50, 177
Photographs, 12
Pickup campers, 18–20
Pilots, 50
Pliers, 44
Plumbing systems (RVs), 28, 29
Poisonous plants, 154, 177–178
Ponchos, 48, 92, 93, 184
Porcupines, 174–175
Pots and pans. *See* Cooking, utensils
Preparation
    check list, 50, 54–55
    for first-timers, 32–33
    travel information, 55, 97–98
Primitive camping, 102, 204–205
Pulley, 49

R

Radios, 32, 35, 50, 188
Rain fly, 58, 59, 60, 61, 65, 92, 101
Rain gear, 39, 48, 85, 90–93, 176, 184–185
Recipes, 132–139
Recreational vehicle camping
    hookups, 23, 203–204
    trip preparation, 54–55, 98
    trips, 198–199
Recreational Vehicle Industry Association (RVIA), 27, 29
Recreational vehicles (RVs)
    driving, 30–31
    emergency equipment for, 50
    features of, 7–8, 18, 21, 30
    fuel consumption, 17, 20, 27–28
    gross vehicle weight, 21, 22, 27
    overloading, 27
    safety tips, 28–31
    sanitary systems, 8, 28–29
    types of, 17–24
    wattage draws, 26–27
Recreation gear, 47, 55
Resources
    addresses, 205–213
    government campgrounds, 98–99, 203–205
    maps, 46, 55, 98, 146, 213
    National Forests, 97, 207
    National Parks, 98, 205, 206–207

private campgrounds, 203–205
state campgrounds, 207–212
Rifle, 47
River trips, 196–197
Ropes, 49, 51, 69, 154–157
Rucksack, 48

S

Safety guidelines
    campfires, 100–101, 102–103, 104–105, 111, 122
    heaters and stoves in tents, 58, 120, 186
    knives, axes, saws, 159–165
    RVs, 29
    tents, 58, 67, 100
    towing and hitches, 24, 25
    water, 170, 189
    winter camping, 188–189
Sanitary systems, 28–29
Saw, 47, 161
Scorpions, 173
Sharpening tool, 46, 47
Shirts, 88, 93
Shock, 175–176
Shock cord, 69
Shovel, 48, 174
Ski touring, 74
Skunks, 174
Sleeping bags
    care of, 74–75, 79
    closures, 73, 79
    construction, 76–79
    insulation, 71–72
    liners, 183, 192
    materials, 39, 48, 76, 85
    parts of, 73
    renting, 58
    selection, 48, 72, 74–79
    shape, 48, 74–76, 188
    for spring and fall, 183
    storage of, 74–75, 78–79
    weather and, 74, 76
    weights, 65
    for winter, 186, 187–188, 192
Slingshot, 49
"S" link, 49
Snack items, 139, 185–186
Snakebite, 168–169
Snow, 187, 189
Snowshoeing, 187
Socks, 86
"Space blanket," 8, 50
Spiders, 172–173
Spring camping, 183–186
Stars, 144, 146
    constellation charts, 150–151

State parks, 204, 207–212
Storage
    in RVs, 18, 22
    sleeping bags, 17, 74–75
    tents, 65–67
Storms, 99, 154
Stoves, 16, 39, 102, 109, 116
    major categories, 118–121
    safety, 111, 122
    in tents, 58, 120, 186
Stretchers, 175
Summer weather chart, 181
Sunglasses, 46
Sunstroke, 170–171
Survival skills, 49–50, 157

T

Tacoma Mountain Search and Rescue Council, 50
Tarp, 67, 69, 186
Tent construction
    closures, 63, 67, 186
    guy lines, 64, 65
    mesh screening, 58, 63
    poles, 7, 61, 64, 67, 69, 102, 187
    rain fly, 58, 59, 60, 61, 65, 92, 101
    tarp, 67, 69, 186
Tent designs, 43, 48, 186
    A-frame or "pup," 59, 61, 62
    baker, 65
    dome, 60, 63, 64
    other shelters, 69
    pyramid or "tepee," 60, 61–64
    tube, 49, 69
    umbrella, 62
    wall, 57, 60, 64–65
Tent(s), 16
    campfires and, 100
    care of, 42, 65–67
    children and, 59
    damage and wear, 58, 59, 61, 65–67, 176, 187
    do-it-yourself, 60
    heaters and stoves in, 58, 120, 121–122, 186
    materials, 7, 39, 43, 48, 60–61, 64
    pitching, 101, 109, 111
    pretesting, 63, 66
    renting, 58–59, 204
    safety, 58, 67, 100, 186
    selecting, 57, 58–59, 65
    size of, 65
    for spring and fall, 186
    weight of, 7, 57–58, 59, 60, 65
    for winter, 186, 187
Tent trailers. *See* Camping trailers

Thinsulate, 71, 72
Ticks, 171–172
Toilets
    "cat," 107, 109
    portable, 107, 108
    for RVs, 28–29
Topographical maps, 46, 47, 145, 146–148, 152, 213
Towing
    hitches, 24–28
    smaller vehicles, 18
    tips, 24
    travel trailers, 21–22
    and weight class, 24
Trailers
    camping, 22–24
    classes, 24
    fifth-wheel, 25–26
    travel, 21–22
    towing tips, 24–28
Trails
    Appalachian Trail, 195
    Chilkoot Trail, 194–195
    Lewis and Clark Trail, 198–199
    in Olympic National Park, 193
    Pacific Crest Trail, 195–196
Trapper blankets, 48
Travel information, preparing, 55, 97–98. *See also* Resources
Travel trailers, 21–22

U
Undergarments, 88

V
Vacuum bottle, 48, 130
Vandalism, 35
Vans, 19, 20, 64
Voyagers National Park, 198

W
Waste disposal. *See* Latrines; Sanitary systems; Toilets
Watch compass, 153–154
Water
    at campgrounds, 204
    carrying, 48
    purifying, 118, 193
    safety, 170, 189
    sources of, 187
Weather
    and choice of clothing, 83, 178, 182, 188
    and choice of sleeping bag, 74, 76, 188
    postponing trip for, 15, 189
    sudden changes in, 99, 178, 189
    summer, chart of, 181
    wind-chill factor, 49, 181
Wet, getting, 49, 178, 188–189
Whistle, 49, 162
Whitewater trips, 196–197
Wildcrafting, 165
Wilderness camping, 102, 191
Wild foods, 139–141
Winch, portable, 174
Windbreakers, 85, 90
Wind-chill factor, 49, 181
Winter camping, 186–189, 192

Y
Yellowstone National Park, 198
Yosemite National Park, 173